СПЕКТРЫ и АНАЛИЗ

SPEKTRY I ANALIZ

SPECTRA AND ANALYSIS

SPECTRA AND ANALYSIS

A. A. Kharkevich

TRANSLATED FROM RUSSIAN

Springer Science+Business Media, LLC 1960

The underlying Russian text
is that of the revised third edition,
published by the State Press for
Technical and Theoretical Literature,
Moscow, in 1957.

ISBN 978-1-4899-4865-6 ISBN 978-1-4899-4863-2 (eBook)
DOI 10.1007/978-1-4899-4863-2

Library of Congress Catalog Card Number: 60-9256

Originally published by Consultants Bureau Enterprises, Inc in 1960.
Softcover reprint of the hardcover 1st edition 1960
227 West 17th St., New York 11, N. Y.

CONTENTS

CHAPTER I – SPECTRA

CHAPTER II – ANALYSIS

CHAPTER III – THE SPECTRA OF RANDOM PROCESSES

APPENDIXES

CHAPTER I

SPECTRA

1. Introduction

In the days when Bernoulli and Euler, followed by Fourier, were first investigating the expansion of functions in trigonometric series,[1] this type of expansion was then considered merely as a mathematical instrument for the solution of the problems of mathematical physics. Fourier himself made use of the series bearing his name for the integration of the equations of thermal conduction. The Fourier method has become a classic example of the solution of wave equations — the string equation and, later, the telegraph equation. However, for some time the Fourier expansion was not directly associated with any physical ideas. Even after the discovery of electrical oscillations and waves there was doubt expressed as to the adequacy of the Fourier expansion as applied to actual physical effects. For example, Hertz (see his correspondence with Poincaré)[2] made a pessimistic allusion to the idea of spectral representations.

For a long time spectral representations were applied and developed only by a relatively small circle of theoretical physicists. Then, beginning with the twenties, in connection with the vigorous growth of electronics, acoustics, vibrational mechanics, and all the branches of engineering in general that are in some way connected with the theory of oscillations, spectral representations found an extraordinarily broad application. The direct connection between a spectral expansion and the behavior of real oscillating systems was established. The spectral method for describing physical effects gained universal recognition.

Moreover, the language of spectra has become a universal language, serving as the medium of communication between all those having anything to do

[1]It is a curious fact that even the possibility of such an expansion was disputed at the time.
[2]Cited in a paper by N. N. Andreev [2].

with the technological applications of the various kinds of oscillations. The language of spectra has become the medium for describing not only physical effects, but the properties of the apparatus as well.

There is no doubt that the extensive development of spectral representations has played a powerful role in the march of science; thanks to it, complex oscillatory effects have become common knowledge to the great majority of technicians and physicists.

Still, the history of the development of spectral representations shows that these representations have sometimes "missed the mark." Gross errors have been fostered and continue to appear. Intense and prolonged discussions have arisen on the most fundamental issues (e.g., the discussion of the side bands in radio transmission), and strange misconceptions have appeared (e.g., the incorrect idea of the bandwidth associated with frequency modulation). Paradoxes of one sort or another have been disclosed. But paradoxes, as the late L. I. Mandel'shtam so wisely put it, are possible only wherever there is not complete understanding, "second-order insight," as he expressed it (unless one is speaking of paradoxes due to an imperfection in the theory itself).

The fact still stands, however, that the spectral approach is faultless. It will never lead to error if it is applied with understanding. The errors and false conceptions mentioned above are not the result of a flaw in the method, but of its injudicious application.

Many of the errors can be avoided if we do not restrict our fullest outlook by spectral limitations, but rather modify, add to, and deepen the scope of spectral theory with the passage of time.

It is of the utmost interest to trace the evolution of spectral representations in more recent years. The primary definition of a spectrum is based on the Fourier transform: integration with respect to time carried out between infinite limits. Thus a time function is subjected to transformation as a whole; the result of the transformation, i.e., the spectrum, depends only on the frequency. However, when real experimental conditions are taken into account one is compelled to introduce a new concept, the concept of the "running spectrum." A running spectrum is defined as the result of a Fourier transformation, but with a variable upper limit of integration and in which time appears as a factor. Thus we encounter a spectral function that depends not only a frequency, but on time as well; this then is an intermediate concept, one which brings frequency and time respresentations nearer to one another. This merging process is further continued: the unquestionably useful concept of the "instantaneous spectrum" is introduced. There remains but one step from the instantaneous spectrum to the instantaneous frequency, after which we can once again speak of a "sinusoid with variable frequency," i.e., we can restore to its rightful position a concept that was very highly esteemed in its day. Thus spectral representation, having encompassed a wide range of problems in the course of its development, returns almost to its original place, but on a significantly higher plane: all the fundamental concepts plus a whole series of intermediate ones are clearly defined and form in the aggregate a powerful and profound research tool.

As new branches of technology continue to evolve, mainly in electronic engineering (e.g., pulse techniques, special forms of modulation, etc.), an ever increasing mastery of spectral ideas is demanded of engineers. Therefore, one of the tasks set forth in the present book is that of presenting the basic problems of the theory of spectra within a framework corresponding, insofar as possible, to the requirements of the times, and at a level somewhat higher than the conceptual level of the ordinary technological course or text.

The first chapter of the book, entitled "Spectra," is devoted to an outline of these problems.

The practical application of spectral representations inevitably leads to the necessity of realizing experimentally the Fourier expansion, i.e., to the harmonic analysis of real phenomena. Although there are in existence a tremendous number of methods for analysis and instrument-analyzers making use of these methods, so far many of the fundamental problems of analysis remain insufficiently developed and, in some instances, not at all clear. In particular, the fundamental requirements imposed on analysis as a measurement process and on the analyzer as a measuring instrument are frequently not established or discussed at all. This can probably be explained by the fact that posing such problems is connected with familiar complications. However, these complications are surmountable, and it remains necessary to try and put the problems of analysis into some logical order. Such an attempt has been made in the second chapter, "Analysis."

The problems associated with spectra of random processes are treated in a separate chapter, the third chapter.

A number of additional problems relating to varied special interests and having direct bearing upon the subject at hand are considered in the "Appendixes."

2. Fourier Series and the Fourier Integral

The concept of the Fourier expansion is considered to be generally known. Therefore, here we shall mention only the fundamental relations and definitions. The reader can find the mathematical details in any textbook.

We shall begin with the definition of a periodic function:

$$f(t) = f(t + nT). \tag{2.1}$$

Here T is a constant quantity and is called the period, n is any integer, positive or negative. The definition (2.1) expresses the fundamental property of a periodic function, namely, the fact that the behavior of a phenomenon is periodically repeated and that the periodicity continues for all time, i.e., over the entire range from $-\infty$ to $+\infty$.

From this we can immediately conclude that periodic effects in the rigorous sense of the definition (2.1) do not and cannot exist in reality. A periodic function is useful as a mathematical abstraction; its relationship with real phenomena will appear later.

Every periodic function – with mathematical limitations that we need not bother about[1] – can be represented by a series of trigonometric functions:

$$f(t) = c_0 + \sum_{k=1}^{\infty} c_k \cos\left(2\pi k \frac{t}{T} - \varphi_k\right). \tag{2.2}$$

The periodic function f(t) thus appears as a summation of components of the form

$$c_k \cos\left(2\pi k \frac{t}{T} - \varphi_k\right),$$

each of which is a sinusoidal oscillation with amplitude c_k and initial phase φ_k. The values of c_k and φ_k should be properly chosen, in such a way that Eq. (2.2) is satisfied. The frequencies of the oscillations which make up the periodic function f(t) form a harmonic sequence; this means that the frequencies of all the components are multiples of the fundamental frequency $\frac{1}{T}$. The individual components are called harmonics. The oscillation with the

[1]Namely, the following: the function is assumed to be bounded, piecewise continuous, and to have over the interval of one period a finite number of extremal values (Dirichlet condition).

frequency $\frac{1}{T}$ is called the first harmonic (k = 1), that with frequency $\frac{2}{T}$ is called the second harmonic (k = 2), and so on.

Equation (2.2) can be rewritten in another, very common form:

$$f(t) = c_0 + \sum_{k=1}^{\infty}\left(a_k \cos 2\pi k \frac{t}{T} + b_k \sin 2\pi k \frac{t}{T}\right), \tag{2.3}$$

where

$$a_k = c_k \cos\varphi_k, \qquad b_k = c_k \sin\varphi_k,$$

so that

$$c_k = \sqrt{a_k^2 + b_k^2}, \qquad \operatorname{tg}\varphi_k = \frac{b_k}{a_k}.$$

The coefficients a_k and b_k are defined by the equations

$$a_k = \frac{2}{T}\int_{-\frac{T}{2}}^{\frac{T}{2}} f(t)\cos 2\pi k \frac{t}{T}\,dt, \tag{2.4}$$

$$b_k = \frac{2}{T}\int_{-\frac{T}{2}}^{\frac{T}{2}} f(t)\sin 2\pi k \frac{t}{T}\,dt. \tag{2.5}$$

The quantity c_0 expresses the mean value of the function over one period; it is usually called the constant component and is computed from the equation

$$c_0 = \frac{1}{T}\int_{-\frac{T}{2}}^{\frac{T}{2}} f(t)\,dt. \tag{2.6}$$

When a_k, b_k, and c_0 are defined by Eqs. (2.4), (2.5), and (2.6), Eq. (2.3) is an identity.

The outstanding property of the Fourier series is the fact that if we take a finite number of terms of the series, i.e., approximate the periodic function by a trigonometric polynomial, writing it in the form

$$f(t) \simeq c_0 + \sum_{k=1}^{N}\left(a_k \cos 2\pi k \frac{t}{T} + b_k \sin 2\pi k \frac{t}{T}\right),$$

then for any value of N the least-square deviation from the exact value is obtained when the coefficients of the polynomial are defined by the same equations (2.4), (2.5), and (2.6). As the number of terms N increases, the approximation, of course, becomes better, and in the limit, as $N \to \infty$, the approximate equation goes over to an exact one.

The Fourier series can also be written in complex form, as follows:

$$f(t) = \sum_{k=-\infty}^{\infty} C_k e^{j2\pi k \frac{t}{T}}, \tag{2.7}$$

where

$$2C_k = c_k e^{-j\varphi_k} = a_k - jb_k; \quad c_k = 2|C_k|; \quad c_0 = C_0.$$

The quantity $2C_k$ is the complex amplitude; C_k is found from the equation

$$C_k = \frac{1}{T}\int_{-\frac{T}{2}}^{\frac{T}{2}} f(t) e^{-j2\pi k \frac{t}{T}}\, dt. \tag{2.8}$$

The summation in (2.7) is taken over every integer k, both positive and negative, and including zero. In order to go from (2.7) to (2.2) or (2.3) it is necessary to recall that the real part of each component under the summation sign in (2.7) is even with respect to k, while the imaginary part is odd. In all that follows we shall use the compact form (2.7) predominantly.

The Fourier series gives an expansion of a periodic function over trigonometric functions. This expansion can be generalized to the case of a nonperiodic function. A nonrigorous, but easily visualized approach to the generation of the Fourier expansion of a nonperiodic function consists in the application of the limiting transition as $T \to \infty$. In fact, a nonperiodic function can be treated as the limiting case of a periodic function whose period is increasing without bound. Let us take Eq. (2.7) and put into it the value of C_k from (2.8):

$$f(t)=\frac{1}{T}\sum_{k=-\infty}^{\infty} e^{j2\pi k\frac{t}{T}}\int_{-\frac{T}{2}}^{\frac{T}{2}} f(t)e^{-j2\pi k\frac{t}{T}}\,dt.$$

Let us pass to the limit, letting T go to infinity. In place of $\frac{1}{T}$ let us introduce the fundamental circular frequency.

$$\omega_1=\frac{2\pi}{T}.$$

This quantity is the frequency inverval between adjacent harmonics, the frequencies of which are equal to $2\pi\frac{k}{T}$. In the limiting transition we shall make the following transformations:

$$T\to\infty,$$
$$\omega_1\to d\omega,$$
$$2\pi\frac{k}{T}\to\omega,$$

where ω is the running frequency, which varies continuously, and $d\omega$ is its increment. The sum goes over to an integral, and we obtain

$$f(t)=\frac{1}{2\pi}\int_{-\infty}^{\infty} e^{j\omega t}\,d\omega\int_{-\infty}^{\infty} f(t)\,e^{-j\omega t}\,dt, \tag{2.9}$$

or

$$f(t)=\frac{1}{2\pi}\int_{-\infty}^{\infty} S(\omega)e^{j\omega t}d\omega, \tag{2.10}$$

where

$$S(\omega)=\int_{-\infty}^{\infty} f(t)\,e^{-j\omega t}\,dt. \tag{2.11}$$

Equations (2.10) and (2.11) are the basic equations of the theory of spectra. They represent a pair of Fourier transformations, connecting two functions, one to the other: the real time function f(t) and the complex function S(ω). Eq. (2.10) represents the Fourier integral in complex form. The meaning of this equation is contained in the fact that the function f(t) is represented by a summation of sinusoidal components. But the function f(t) is assumed to be nonperiodic; therefore, it can be written only as the sum of an infinitely large number of infinitesimal oscillations, which are infinitely close to one another in frequency. The complex amplitude of each individual oscillation is infinitesimal; it is equal to

$$dC = \frac{1}{\pi} S(\omega)\, d\omega. \tag{2.12}$$

The frequency interval between two adjacent oscillations is also infinitesimal; it is equal to $d\omega$.

If the Fourier series represents a periodic function by a summation of an infinite number of sinusoids, but with frequencies having definite discrete values, then the Fourier integral represents a nonperiodic function by a summation of sinusoids with a continuous sequence of frequencies. In the composition of a nonperiodic function there appear, so to speak, all frequencies.

One of the features that distinguishes the Fourier integral from the Fourier series is the fact that the Fourier series represents a periodic function as the summation of other periodic components, whereas the Fourier integral represents a nonperiodic function as the sum of periodic components. Thus in the case of the Fourier integral the summation does not have the same essential properties as its components, a fact which must be borne in mind when considering the general nature of a spectral expansion according to Fourier.

Let us note, in conclusion, that Eq. (2.10) can be written in real form; then the integration will be performed only over positive frequencies. Introducing the notation

$$S(\omega) = A(\omega) + jB(\omega),$$

we obtain (remembering that A is an even and B an odd function)

$$f(t) = \frac{1}{\pi} \int_0^\infty [A(\omega) \cos \omega t - B(\omega) \sin \omega t]\, d\omega. \tag{2.13}$$

It is possible to obtain Eq. (2.10) in still another form, by writing as follows:

$$f(t) = \frac{1}{2\pi} \int_0^\infty [S(\omega)\, e^{j\omega t} + S(-\omega)\, e^{-j\omega t}]\, d\omega.$$

In the quantity contained in the brackets we find a summation of conjugates, which is equal to twice the real part. Therefore,

$$f(t)=\frac{1}{\pi} Re \int_0^\infty S(\omega)\, e^{j\omega t}\, d\omega. \tag{2.14}$$

3. Spectra; Definition and Classification

Let us refer to the Fourier series equation (2.2) and rewrite it in the form

$$f(t)=c_0+\sum_{k=1}^{\infty} c_k \cos(k\omega_1 t-\varphi_k).$$

Here $\omega_1 = \dfrac{2\pi}{T}$ is the fundamental frequency. As we can see, the complex periodic function f(t) is fully defined by the set of quantities c_k and φ_k. The set of quantities c_k is called the amplitude spectrum. The set of quantities φ_k is correspondingly called the phase spectrum. For many applications it is sufficient to know the amplitude spectrum alone; it is applied so predominantly that when speaking simply of spectrum, the amplitude spectrum is usually implied. Otherwise appropriate reservations must be made. We shall do the same here.

The spectrum of a periodic function can be illustrated graphically. Let us take for this purpose the coordinates c_k and $\omega = k\omega_1$. The spectrum will be represented in this system of coordinates by a set of discrete points, since each value of $k\omega_1$ will correspond to a single definite value of c_k. A graph consisting of individual points is not quite suitable. Therefore, the amplitudes of the separate harmonics are conventionally represented by vertical segments of appropriate length. As a result, the spectrum of a periodic function assumes the form shown in Fig. 1. This is a discrete spectrum; it is also called a "line" spectrum, borrowing this terminology from optics. The second property of the spectrum shown in Fig. 1 is contained in the fact that the spectrum is harmonic. This means that it consists of equally spaced spectral lines; the frequencies of the harmonics stand in simple multiple proportion to one another. Of course, the individual harmonics, occasionally even the first, can be absent, i.e., their amplitudes can be equal to zero; this, however, does not violate the harmonic character of the spectrum.

It should not be thought that only a periodic function has a discrete spectrum. Let us assume, for example, that a complex oscillation is the result of the compounding of two sinusoidal oscillations with incommensurate frequencies, say, ω_1 and $\sqrt{2}\,\omega_1$. This oscillation is known to be nonperiodic, but its spectrum is discrete and consists of two spectral lines.

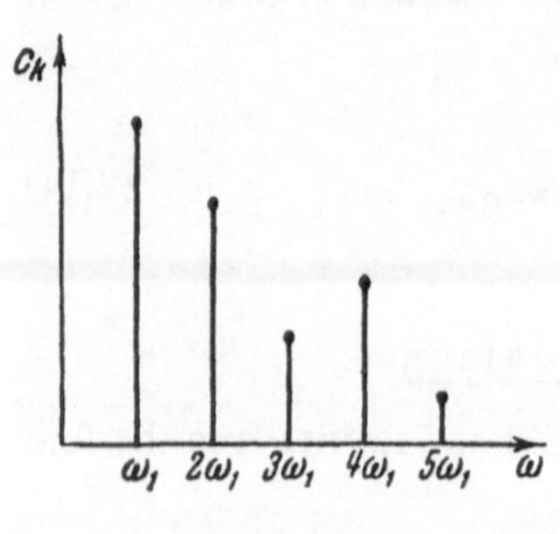

Fig. 1.

A function which possesses a discrete spectrum consisting of spectral lines distributed arbitrarily over a frequency scale is called quasi-periodic and has many interesting properties.[1]

Thus, discrete, or ruled spectra can be ascribed to both periodic and nonperiodic functions. In the former case the line spectrum is harmonic by implication.

The special case of a quasi-periodic function represented by an expansion of the form

$$f(t) = \sum c_k \cos[(\omega_0 + k\omega_1)t - \varphi_k],$$

where k assumes both negative and positive values, is of great practical interest. The spectrum corresponding to this expansion is characterized by the fact that its lines are equidistant; therefore, we shall call this type of line spectrum quasi-harmonic. An example of such spectra is contained in periodically modulated oscillations; ω_0 in this case is nothing more than the carrier frequency.

Let us turn now to the spectra of nonperiodic functions. We know already that, as a result of the limiting transition from the Fourier series to the Fourier integral, the intervals between the individual spectral lines are contracted to infinitesimals, the lines merge together, and, instead of discrete points, the spectrum must be represented by a continuous sequence of points, i.e., by a continuous curve. This kind of spectrum is called a continuous spectrum.

In this case it becomes necessary to introduce one refinement. We wrote the equation for the Fourier integral in the form (2.10):

$$f(t) = \frac{1}{2\pi} \int_{-\infty}^{+\infty} S(\omega) e^{j\omega t} d\omega.$$

The integrand expresses a separate infinitesimal component, i.e., an oscillation with infinitesimal amplitude dC:

[1]In this respect, see G. Bor, Quasi-Periodic Functions [in Russian] (GTTI, 1934); a brief outline of the theory of quasi-periodic functions is given in the appendix to the following book: I. I. Prizalov, Fourier Series [in Russian] (GTTI, 1934).

$$\frac{1}{\pi} S(\omega) e^{j\omega t} d\omega = dC e^{j\omega t}.$$

From this we find

$$S(\omega) = \pi \frac{dC}{d\omega}.$$

Thus the quantity $S(\omega)$ does not express the amplitude directly, but, rather, the so-called spectral density. This fine point, however, is usually bypassed and $S(\omega)$ is called the complex spectrum of a nonperiodic function, and the absolute value (modulus) of this quantity,

$$\Phi(\omega) = |S(\omega)|$$

is called, simply, the spectrum. This can lead to misunderstanding only when we wish to equate the relations for periodic and nonperiodic functions directly.

Thus we have two fundamental categories of spectra: line and complex. Harmonic line spectra are ascribed to periodic functions, complex speotra to nonperiodic functions.

In concluding the present section, we note that some time function can be used to express the variation of the most diverse physical quantities. The spectra of these functions then take on a corresponding meaning. In practice it is necessary to be concerned, for example, with the spectra of mechanical quantities: displacement, velocity, acceleration, force, pressure, etc.; electrical quantities: current, potential, etc. Moreover, we are often interested in the spectra of quadratic quantities: power or energy. For example, in optics we are concerned most of all with these quantities.

4. Spectral Theorems

We shall now introduce a few general spectral theorems, based on the properties of the Fourier transform. These theorems are similar to the theorems of operational calculus and are derived in an analogous manner: the Fourier transform and the Laplace transform, which form the basis for the operational calculus, are closely related to one another.

We shall first note that the Fourier transform is linear. From this it follows immediately that we can apply the principle of superposition to it. This fact can be expressed by the following relation:

$$S_1(\omega)+S_2(\omega)=\int\limits_{-\infty}^{+\infty} e^{-j\omega t}[f_1(t)+f_2(t)]\,dt. \tag{4.1}$$

The interpretation of Eq. (4.1) can be briefly expressed as follows: the spectrum of the sum is equal to the sum of the spectra.

Let us now verify the theorem of the spectrum of a derivative: if the complex spectrum of a function f (t) is S (ω), then the complex spectrum of $f'(t) = \frac{df}{dt}$ is $j\omega S(\omega)$; the complex spectrum of a derivative is obtained from the complex spectrum of the function, multiplying by $j\omega$.

In order to prove the theorem, let us compose the expression.

$$S_{(1)}(\omega)=\int\limits_{-\infty}^{+\infty} e^{-j\omega t} f'(t)\,dt$$

and integrate it by parts. We obtain

$$S_{(1)}(\omega)=f(t)\,e^{-j\omega t}\Big|_{-\infty}^{+\infty}+j\omega\int\limits_{-\infty}^{+\infty} f(t)\,e^{-j\omega t}\,dt.$$

Since any function that can be represented by a Fourier integral reverts to zero as $t \to \pm\infty$, we have

$$S_{(1)}(\omega)=j\omega S(\omega). \tag{4.2}$$

This proof can be extended to the case of the nth derivative. Performing the integration by parts n times, we obtain the complex spectrum of the nth derivative [under the condition that all derivatives of the function up to the $(n-1)$th order inclusive go to zero as $t \to \pm\infty$]:

$$S_{(n)}(\omega)=(j\omega)^n S(\omega). \tag{4.3}$$

An expression for the complex spectrum of the integral of the given function can be derived in exactly the same way.

Writing the expression

$$S_{(-1)}(\omega)=\int\limits_{-\infty}^{+\infty} e^{-j\omega t}\left(\int\limits_{-\infty}^{t} f(\tau)\,d\tau\right)dt$$

and integrating it by parts, we find

$$S_{(-1)}(\omega) = \frac{1}{j\omega} S(\omega) \tag{4.4}$$

under the condition that

$$\int_{-\infty}^{+\infty} f(t)\,dt = 0.$$

This condition is satisfied, for example, for all odd functions integrated within infinite limits.

Let us now derive an expression for the complex spectrum of a function differing from the original function by a time delay τ. We can write

$$S_{\tau}(\omega) = \int_{-\infty}^{+\infty} e^{-j\omega t} f(t-\tau)\,dt.$$

By means of a simple substitution of the variable according to the equation $t_1 = t - \tau$ we arrive at the result

$$S_{\tau}(\omega) = e^{-j\omega\tau} S(\omega). \tag{4.5}$$

If in this relation we go from complex spectra to their absolute values, we obtain

$$|S_{\tau}(\omega)| = \Phi_{\tau}(\omega) = \Phi(\omega),$$

i.e., for a time delay, or, in general, for a displacement of the function along the time scale, the spectrum does not depend, as one might expect on the choice of initial time of reference.

The next theorem refers to the transposition of spectra. The problem is posed as follows: what kind of function will correspond to a spectrum shifted along the frequency scale by the amount Ω?

Since

$$S(\omega + \Omega) = \int_{-\infty}^{+\infty} e^{-j(\omega+\Omega)t} f(t)\,dt,$$

is a consequence, the complex spectrum of the form we are seeking will be ascribed to the function

$$f_2(t) = e^{-j\Omega t} f(t). \tag{4.6}$$

Let us now derive a few more complicated relations. We shall take the expression for the Fourier integral:

$$f_1(t) = \frac{1}{2\pi} \int_{-\infty}^{+\infty} S_1(\omega)\, e^{j\omega t}\, d\omega,$$

multiply both sides by $f_2(t)$, and integrate over t within the limits $\pm\infty$. We obtain

$$\int_{-\infty}^{+\infty} f_1(t) f_2(t)\, dt = \frac{1}{2\pi} \int_{-\infty}^{+\infty} f_2(t)\, dt \int_{-\infty}^{+\infty} S_1(\omega)\, e^{j\omega t}\, d\omega.$$

Let us invert the order of integration on the right-hand side:

$$\int_{-\infty}^{+\infty} f_2(t)\, dt \int_{-\infty}^{+\infty} S_1(\omega)\, e^{j\omega t}\, d\omega = \int_{-\infty}^{+\infty} S_1(\omega)\, d\omega \int_{-\infty}^{+\infty} f_2(t)\, e^{j\omega t}\, dt.$$

Then

$$\int_{-\infty}^{+\infty} f_1(t) f_2(t)\, dt = \frac{1}{2\pi} \int_{-\infty}^{+\infty} S_1(\omega)\, S_2(-\omega)\, d\omega. \tag{4.7}$$

This equation is convenient, for example, in calculating the energy when the spectra of the current and potential are known, or for any two functions whose product expresses power.

In real form Eq. (4.7) can be written as follows [taking into account the fact that $S(-\omega) = S^*(\omega)$];

$$\int_{-\infty}^{+\infty} f_1(t) f_2(t)\, dt = \frac{1}{\pi} \int_{0}^{\infty} \Phi_1(\omega)\, \Phi_2(\omega) \cos(\varphi_1 - \varphi_2)\, d\omega; \tag{4.8}$$

for the special case $f_1 = f_2$.

$$\int_{-\infty}^{+\infty} f^2(t)\,dt = \frac{1}{\pi}\int_{0}^{\infty} \Phi^2(\omega)\,d\omega. \tag{4.9}$$

The last relation is well known as the Rayleigh theorem. There will be considerable occasion for us to make use of it later. Equation (4.9) shows that in the physical sense the function

$$\Phi^2(\omega) = S(\omega)\,S^*(\omega)$$

represents the spectral energy density.

Let us rewrite Eq. (4.7) in the new notation

$$\int_{-\infty}^{+\infty} f_1(t)\,u_2(t)\,dt = \frac{1}{2\pi}\int_{-\infty}^{+\infty} S_1(\nu)\,R_2(-\nu)\,d\nu$$

and define the function $u_2(t)$ by the relation

$$u_2(t) = f_2(t)\,e^{-j\omega t}.$$

If S_2 denotes the spectrum of the function f_2, then, on the basis of the theorem (4.6) we have

$$R_2(\nu) = S_2(\omega + \nu)$$

and, consequently,

$$\int_{-\infty}^{+\infty} f_1(t)\,f_2(t)\,e^{-j\omega t}\,dt = \frac{1}{2\pi}\int_{-\infty}^{+\infty} S_1(\nu)\,S_2(\omega - \nu)\,d\nu.$$

Thus if S_1 and S_2 are the spectra of the functions f_1 and f_2, respectively, while S is the spectrum of the product $f = f_1 f_2$, then we have

$$S(\omega) = \frac{1}{2\pi}\int_{-\infty}^{+\infty} S_1(\nu)\,S_2(\omega - \nu)\,d\nu. \tag{4.10}$$

The integral in the right-hand side is called the "packet" of the functions S_1 and S_2.

Equation (4.10) expresses the spectrum of a product of two functions in terms of the spectra of each.

Finally, we shall derive one further equation. Let us write the packet of two time functions f_1 and f_2:

$$f(t)=\int_{-\infty}^{+\infty} f_1(\tau) f_2(t-\tau)\,d\tau$$

and compute the spectrum of this function

$$\begin{aligned}S(\omega)&=\int_{-\infty}^{+\infty} e^{-j\omega t}\,dt\int_{-\infty}^{+\infty} f_1(\tau) f_2(t-\tau)\,d\tau=\\&=\int_{-\infty}^{+\infty} f_1(\tau)\,d\tau\int_{-\infty}^{+\infty} e^{-j\omega t} f_2(t-\tau)\,dt=\\&=\int_{-\infty}^{+\infty} e^{-j\omega t} f_1(\tau)\,d\tau\int_{-\infty}^{+\infty} e^{-j\omega\mu} f_2(\mu)\,d\mu.\end{aligned}$$

Here, after inverting the order of integration, a change of variable is made according to the equation $\mu = t-\tau$.

Thus the spectrum of the function $f(t)$ is

$$S(\omega)=S_1(\omega)\,S_2(\omega). \tag{4.11}$$

This relation makes it possible to find a time function whose spectrum is known and which expresses the product of the spectra of two functions. Assuming in (4.11) that

$$S_2(\omega)=S_1^*(\omega),$$

we can be sure that $\Phi^2(\omega)$ is the spectrum of the function

$$\int_{-\infty}^{\infty} f(\tau) f(t+\tau)\,d\tau=\int_{-\infty}^{\infty} f(\tau) f(\tau-t)\,d\tau.$$

The application of Eqs. (4.1)-(4.11) will simplify to a considerable extent the computation of the spectra of various functions.

In concluding the present section, one interesting fact must be pointed out. By applying the Fourier expansion, we are dealing with a pair of Fourier transforms:

$$\left.\begin{aligned} S(\omega) &= \int_{-\infty}^{+\infty} f(t)\, e^{-j\omega t}\, dt, \\ f(t) &= \frac{1}{2\pi}\int_{-\infty}^{+\infty} S(\omega)\, e^{j\omega t}\, d\omega. \end{aligned}\right\} \tag{4.12}$$

In these equations attention is drawn to the fact that the time t and circular frequency ω appear here symmetrically, with equal origins.[1] But from the complete symmetry of Eq. (4.12) it also follows that every theorem in the theory of spectra has a counter theorem, requiring no further proof and obtained from the first theorem by a simple formal change of the variable t to the variable ω and of the time functions to the corresponding spectral densities.

Looking back over the theorems of the present section, we are convinced of the fact that they have a paired nature. In order to illustrate this, we can write down some of them in the form of a table:

Original transformations	
$S(\omega) = \int_{-\infty}^{+\infty} f(t)\, e^{-j\omega t}\, dt$	$f(t) = \frac{1}{2\pi}\int_{-\infty}^{+\infty} S(\omega)\, e^{j\omega t}\, d\omega$
$S_\tau = \int_{-\infty}^{+\infty} f(t-\tau)\, e^{-j\omega t}\, dt =$ $= e^{-j\omega\tau} S$ Time-delay theorem	$f_\Omega = \frac{1}{2\pi}\int_{-\infty}^{+\infty} S(\omega+\Omega)\, e^{j\omega t}\, d\omega =$ $= e^{-j\Omega t} f$ Displacement theorem
$S = \int_{-\infty}^{+\infty} f_1 f_2 e^{-j\omega t}\, dt =$ $= \frac{1}{2\pi}\int_{-\infty}^{+\infty} S_1(\nu)\, S_2(\omega-\nu)\, d\nu$ Theorem of the spectrum of a product	$f = \frac{1}{2\pi}\int_{-\infty}^{+\infty} S_1 S_2 e^{j\omega t}\, d\omega =$ $= \int_{-\infty}^{+\infty} f_1(\tau)\, f_2(t-\tau)\, d\tau$ Theorem of the spectrum of a packet

[1] The equation can be made perfectly symmetrical by modifying the original definition in such a way that the factor $1/2\pi$ is shared by both integrals (i.e., introducing the factor $1/\sqrt{2\pi}$ into both equations, a procedure which is customarily carried out.

5. Running Spectra

From the fundamental definition (Sections 2 and 3) the spectral density is expressed by the equation

$$S(\omega) = \int_{-\infty}^{+\infty} f(t)\, e^{-j\omega t}\, dt. \tag{5.1}$$

Thus, in order to determine a spectrum, it is necessary to perform an integration with respect to time within infinite limits. This is possible in principle when the function f(t) is specified and known over the entire infinite range of the time axis. But if the function f(t) stands for some real physical process, which is an object of our observation, and if the entire behavior of this process cannot be predicted exactly on the basis of theoretical considerations, then the information concerning the function f(t) is obtained only as the result of our observations. Therefore, we cannot perform the integration over an infinite range as required by the definition (5.1), but only up to and including the present, and ever changing, time.

Everything that has already passed can in principle be known to us, so that the integration can be carried out within the limits from $-\infty$ to the variable time _t_. With this modification the definition of the spectrum assumes the form

$$S_t(\omega) = \int_{-\infty}^{t} f(t)\, e^{-j\omega t}\, dt. \tag{5.2}$$

The quantity $S_t(\omega)$ is then a function, not only of the frequency, but also of time, and is called the running spectrum.

Under real conditions the observation of a process (or the process itself) can actually be started at some instant t_0, which is displaced into the past at some finite interval from the passing time _t_. In this case the instant t_0 can be taken as the time reference, and we can define the running spectrum as follows:

$$S_t(\omega) = \int_{0}^{t} f(\tau)\, e^{-j\omega t}\, d\tau. \tag{5.3}$$

In what follows we shall use both definitions of the running spectrum.

It is clear that an expression which defines a spectrum mathematically under the conditions of a real experiment is by its very existence of the utmost significance. This side of the question is discussed in more detail in Section

18. But the concept of the running spectrum is, in general, a very fruitful one.

In the very beginning of the exposition of the theory of spectra we discussed the spectrum of a periodic function, defined by the relation

$$f(t) = f(t + nT). \tag{2.1}$$

A periodic function is a mathematical abstraction. This abstraction is very useful. But it must be borne in mind that there cannot exist any sort of real process corresponding to the definition (2.1). Every real process has a beginning and an end, and, consequently, is described by an expression of the form (2.1) over some finite interval of time. We shall consider a real cyclic process periodic when this process continues for a sufficiently long period of time. The number of "periods" will serve as the gage of the continuance; the continuance will be large whenever the number of periods is many times greater than unity. If we select a short segment of some process, then it will not have a periodic character in the fullest sense. The periodicity of a process does not manifest itself at one stroke; only with the passage of time do the characteristic features of a process become evident. The running spectrum expresses just this evolution of a process from the spectral point of view.

The spectrum of a short interval during a process—over a short period of time from its initiation—is homogeneous, since a short segment of any process is simply a short pulse. If at a later time the periodic repetition of a certain cycle of an effect occurs, then maxima will begin to form at the fundamental frequency and its harmonics in the running spectrum. These maxima become sharper and higher, while the value of the spectral density in the intervals between the maxima diminishes and only in the limit, as $t \to \infty$, does the continuous running spectrum degenerate into the line spectrum of a process that is periodic in the strict sense of the word.

Naturally, for a process over a sufficiently long period of time the maxima will be made so narrow that they may, for all practical purposes, be treated as lines. However, this does not in principle detract from the significance of all that has been stated above — a periodic process is only a limit to which an actual repetitive process can tend with the passage of time.

In order to clear up some of these considerations, let us construct the running spectrum of a sinusoid. Applying the definition (5.3) and substituting the expression

$$f(t) = \sin \Omega t,$$

into it, we find

$$S_t = \int_0^t e^{-j\omega t} \sin \Omega t \, dt| = \frac{\Omega}{\Omega^2 - \omega^2}\left[1 - e^{-j\omega t} \left(\cos \Omega t + j\frac{\omega}{\Omega} \sin \Omega t\right)\right]. \tag{5.4}$$

Equation (5.4) can be greatly simplified by considering the value of the spectral density for the discrete times

$$t = t_n = n\frac{T}{2} = n\frac{\pi}{\Omega}.$$

Substituting this value into (5.4), we obtain

$$S_t = \frac{1}{\Omega}\frac{1}{1-\left(\frac{\omega}{\Omega}\right)^2}\left[1-(-1)^n e^{-jn\pi\frac{\omega}{\Omega}}\right]$$

and the spectrum

$$\Phi_t = |S_t| = \frac{2}{\Omega}\frac{1}{1-\left(\frac{\omega}{\Omega}\right)^2} \begin{matrix}\sin \\ \cos\end{matrix}\, n\frac{\pi}{2}\frac{\omega}{\Omega}. \tag{5.5}$$

In this equation the sine refers to even integers $\underline{n}$, the cosine to odd values. The quantity $\underline{n}$ designates the number of half periods of the sinusoid since the moment of initiation.

The indeterminateness at $\omega = \Omega$ is easily resolved:

$$\Phi_t|_{\omega=\Omega} = \frac{1}{2}t = n\frac{T}{4},$$

i.e., the spectral density at this frequency increases linearly with time.

The running spectrum of a sinusoid, as computed from Eq. (5.5), is shown in Fig. 2 in the form of a relief diagram. Along the horizontal axis lying in the plane of the drawing, the frequency ratio $\frac{\omega}{\Omega}$ is plotted, the spectral density is plotted along the ordinate axis, and the number of half periods $\underline{n}$ is plotted along the horizontal axis directed away from the observer. This number is obviously proportional to the time. The details of the left-hand slope of the relief are omitted in order not to encumber the drawing.

Figure 2 clearly demonstrates that the beginning of the spectrum is uniform; the maximum at the frequency Ω is formed very gradually; with the passage of

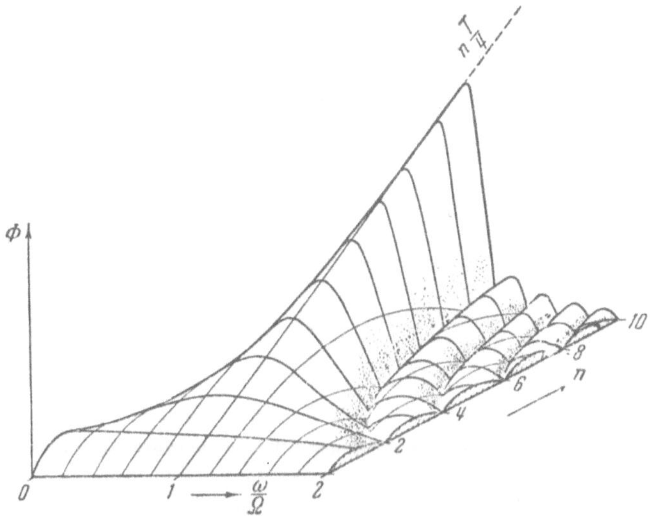

Fig. 2.

time this maximum becomes sharper and sharper, but only in the limit as $t \to \infty$ does the figure go over to a discrete spectral line, which represents a periodic sinusoidal oscillation. Here the spectral density at the frequency $\omega = \Omega$ will be infinitely large. This is as it should be. In this respect, it is necessary to remember what was said in Section 4 about the relationship between the amplitude spectrum of the components of a periodic function and the spectral density of a nonperiodic function.

6. The Instantaneous Spectrum

The introduction of the concept of a running spectrum considerably expands the application of spectral representations. This concept bridges the gap between the frequency and time descriptions of phenomena. But the need is felt for bringing the frequency and time points of view even closer to one another; as a result there arises one more concept: the concept of an instantaneous spectrum.

The ordinary definition of a spectrum (5.1) gives a function of frequency: the time dependence drops out, since the spectrum reflects the process over all time, as a whole. The definition of a running spectrum (5.2) reflects the entire past history (up to the present time) of a process. But it obviously has the meaning and interpretation of a time-variable spectrum, one that reflects the properties of the process at a given instant. Let us imagine, for example, a conversation between a man and a woman, i.e., persons speaking, one with a high, the other with a low voice. It is perfectly reasonable to believe that when the man is speaking the sound spectrum will be distributed more in the low-frequency interval, while the woman's voice will generate a spectrum shifted toward the region of higher frequencies. It remains for us to give this intuitively clear idea a distinct mathematical description.

The simplest definition of an instantaneous spectrum can be given in the following form:

$$S_T(\omega, t) = \int_{t-T}^{t} f(\tau) e^{-j\omega\tau} d\tau, \tag{6.1}$$

i.e., the instantaneous spectrum is defined as the spectrum of a segment of a process, where this segment has a duration T and immediately precedes a given instant _t_. In this definition we are concerned with a "sliding" integration: the interval of integration is of constant _length_, but moves along the time axis; the _location_ of the interval is invariant with respect to the passing time _t_.

A more general definition of the instantaneous spectrum is possible if we introduce a sliding (i.e., connected with running time) weighting function into the integrand. The definition then assumes the form

$$S_r(\omega, t) = \int_{-\infty}^{\infty} r(\tau - t) f(\tau) e^{-j\omega\tau} d\tau. \tag{6.2}$$

It is not difficult to see that the definition (6.1) is a special case of (6.2) when the weighting function assumes the form

$$r(x) = \sigma(x + T) - \sigma(x),$$

where

$$\sigma(x) = \begin{cases} 0 & \text{for } x < 0, \\ 1 & \text{for } x > 0, \end{cases}$$

is a unit step function. These relations are illustrated in Fig. 3. Fano [22] adopts the definition (6.2), choosing the weighting function in the form [1]

[1] Here the constant factor $(2\alpha)^{\frac{1}{2}}$, which was of concern to Fano but is of no importance to us, is omitted.

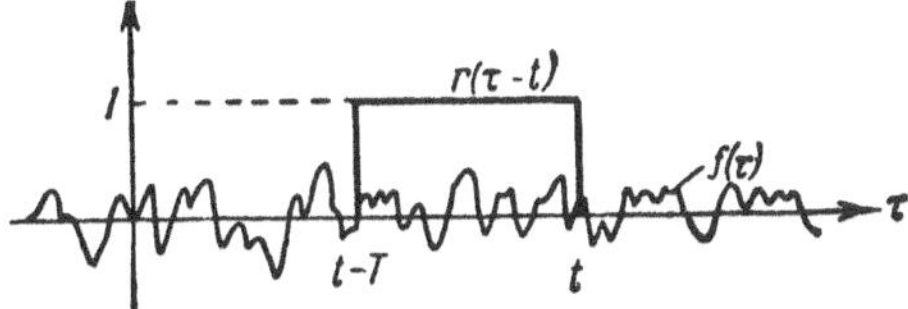

Fig. 3.

$$r(x) = e^{\alpha x}\sigma(-x)$$

(Fig. 4). This function takes into account the entire past history of the process, but weighted exponentially decreasing away from the present time into the past. The basis for such a choice of weighting function is the fact that it expresses the actual result of spectral analysis by means of real filters, characterized by the time constant $\frac{1}{\alpha}$. It will be necessary for us to turn to this problem in Chapter 2.

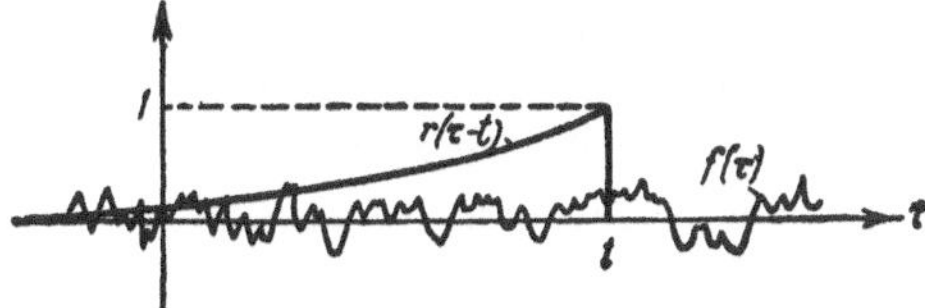

Fig. 4.

Let us turn once again to the definition (6.1) and rewrite it in the form

$$S_T(\omega, t) = \int_{-\infty}^{t} f(\tau) e^{-j\omega\tau} d\tau - \int_{-\infty}^{t-T} f(\tau) e^{-j\omega\tau} d\tau.$$

Thus the instantaneous spectrum represents the difference between two running spectra, or, in other words, the growth in the running spectrum during the time interval T. When T is sufficiently small this growth could be expressed in terms of the derivative of the running spectrum with respect to time

$$\Delta S_t \approx \frac{\partial S_t}{\partial t} T.$$

This consideration brings us to the definition of an instantaneous spectrum according to Page [27]. Page defines the instantaneous spectrum of power as follows:

$$\rho(\omega, t) = \frac{\partial}{\partial t} \left| S_t(\omega) \right|^2, \tag{6.3}$$

where $S_t(\omega)$ is the running spectrum. The integral of the instantaneous power spectrum over the entire frequency axis gives the instantaneous power, i.e., the square of the function f(t):

$$\frac{1}{\pi} \int_0^\infty \rho(\omega, t)\, d\omega = f^2(t).$$

The integral of the instantaneous power spectrum over all past time gives the square of the absolute value of the running spectrum:

$$\int_{-\infty}^{t} \rho(\omega, t)\, dt = \left| S_t(\omega) \right|^2.$$

As we can see, the instantaneous spectrum can be defined in different ways. One thing should not be confused: all definitions are arbitrary. It is only necessary to choose the definition which is convenient for a given instance (for example, one which takes into account the way in which the measuring apparatus behaves) and formulate it clearly from the very beginning, in order to guarantee consistency throughout the discussion. Unfortunately, this elementary requirement is far from being always fulfilled.

7. Spectra of Modulated Oscillations

We shall now proceed from general considerations to a number of applications, being first of all concerned with the problem of modulation, which plays a very important role in modern technology.

Modulation has its greatest application in communications engineering. Every electronically transmitted signal—whether it be a telegraph, telephone, television, or any other signal—is generated by means of modulation. Transmission from a radio station without modulation is like a blank page, whereas modulated transmission is like a page on which all sorts of letters and symbols are printed.

Modulation is of great significance also in modern instrument technique and in a number of specialized branches.

In the transmission of signals a certain physical agent, called the carrier and characterized by a definite number of constant parameters in the absence of modulation, is employed. Modulation is what happens when one or more of the parameters of the carrier is varied with respect to time in correspondence with a delivered signal. In the simplest case, to which we are limiting our discussion, a sinusoidal oscillation is used for the carrier.[1] Let us write the analytic expression for such an oscillation:

$$x = c_0 \sin(\omega_0 t + \varphi_0). \tag{7.1}$$

Here c_0 is the amplitude, ω_0 is the frequency, φ_0 is the initial phase. In the unmodulated oscillation these three parameters, which completely define the oscillation, are constant. In principle it is possible to modulate each of the three indicated parametric quantities; we then have, respectively, amplitude modulation (AM), frequency modulation (FM), and phase modulation (PM). We shall break down each form of modulation in detail.

Let us think of the effect which we call modulation as the multiplication of the modulated quantity by the factor

$$1 + mf(t),$$

where f(t) is the modulating function, defined in such a way that $|f(t)| \leq 1$, and m is the quantity which characterizes the extent of the action and which can assume values from 0 to 1; it is called the modulation depth.

In amplitude modulation the modulated oscillation assumes the form

$$x = c_0[1 + mf(t)] \sin(\omega_0 t + \varphi_0). \tag{7.2}$$

Let us start with the simplest case of a sinusoidal modulation, i.e., let us assume

$$f(t) = \sin \Omega t.$$

Substituting in (7.2), we obtain

$$\begin{aligned} x &= c_0[\sin(\omega_0 t + \varphi_0) + m \sin(\omega_0 t + \varphi_0) \sin \Omega t] = \\ &= c_0 \Big\{ \sin(\omega_0 t + \varphi_0) + \frac{m}{2} \cos[(\omega_0 - \Omega)t + \varphi_0] - \\ &\qquad - \frac{m}{2} \cos[(\omega_0 + \Omega)t + \varphi_0] \Big\}. \end{aligned} \tag{7.3}$$

[1]In the pulse method of transmission a periodic sequence of pulses of one shape or another serves as the carrier.

Consequently, the modulated oscillation has a discrete spectrum consisting of three spectral lines, as shown in Fig. 5. The frequency of the unmodulated oscillation ω_0 is called the carrier frequency; the additional frequencies that arise as a result of the modulation, $\omega_0 - \Omega$ and $\omega_0 + \Omega$, are called the side frequencies, or satellites.

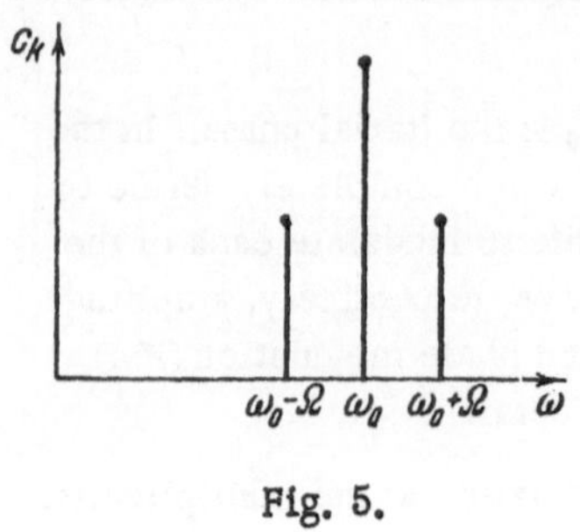

Fig. 5.

Thus an oscillation with a constant frequency, but with a variable amplitude is broken down into several sinusoidal oscillations. We recall that, by definition, only an oscillation of the form (7.1) is sinusoidal, i.e., an oscillation with constant amplitude, frequency, and phase. A modulated sinusoid is then really no longer a sinusoid. This fact can be expressed in other ways, but in the way we have formulated it, it is the key to the proper understanding of the problems of modulation.

Let us consider the more complicated case that arises when the modulating function is periodic, i.e.,

$$f(t) = \sum_{k=1}^{\infty} c_k \sin(k\Omega t + \varphi_k).$$

Then

$$x = c_0\left[1 + m\sum_{k=1}^{\infty} c_k \sin(k\Omega t + \varphi_k)\right]\sin(\omega_0 t + \varphi_0) =$$

$$= c_0\left\{\sin(\omega_0 t + \varphi_0) + \frac{m}{2}\sum_{k=1}^{\infty} c_k \cos[(\omega_0 - k\Omega)t + \varphi_0 - \varphi_k] -\right.$$

$$\left. - \frac{m}{2}\sum_{k=1}^{\infty} c_k \cos[(\omega_0 + k\Omega)t + \varphi_0 + \varphi_k]\right\}.$$

The modulated oscillation consists of the carrier frequency and two groups, called the side bands (both sums contained in the braces). The spectrum of the modulated oscillation is illustrated in Fig. 6.

It should be noted that the right-hand side band reproduces the spectrum of the modulating function, while the left-hand one represents the mirror image of the right-hand one. Thus during the process of modulation a transposition is effected in the spectrum of the modulating function; the spectrum is shifted by the amount ω_0 along the frequency scale. Any general conclusions in this respect can be deduced from the theorem (4.6). We shall have further occasion to do this.

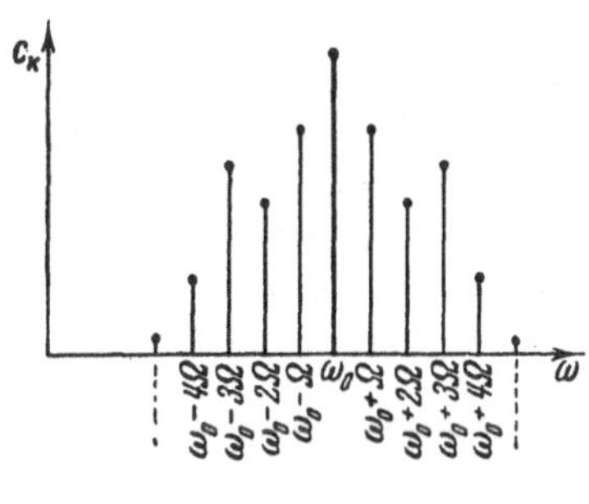

Fig. 6.

It is evident that if the carrier frequency and the fundamental of the modulating function are incommensurate, the resulting oscillation will be nonperiodic and the spectrum shown in Fig. 6 will be quasi-harmonic.

We might add that in the case of so-called balanced modulation there is no carrier frequency, and the spectrum consists only of the side bands. Mathematically this is expressed by the fact that the amplitude of the carrier frequency is multiplied, not by $1 + mf(t)$, but simply by $f(t)$. In fact, a component of the carrier frequency can appear in the modulated spectrum only as the result of a constant component in the expression $1 + mf(t)$. In balanced modulation the oscillation at the carrier frequency is simply multiplied by the modulating function. Let us write the former in the form

$$x = c_0 \cos \omega_0 t = \frac{c_0}{2} \left(e^{j\omega_0 t} + e^{-j\omega_0 t} \right).$$

For the balanced-modulation oscillation we obtain

$$x = \frac{c_0}{2} \left(e^{j\omega_0 t} + e^{-j\omega_0 t} \right) f(t).$$

Applying the theorem (4.6), we find that the spectrum of the modulated oscillation is

$$\frac{c_0}{2} \left[S(\omega_0 - \omega) + S(\omega_0 + \omega) \right],$$

where $S(\omega)$ is the spectrum of the modulating function $f(t)$. The two terms in the brackets express both side bands.

The problem of the width of the modulated spectrum is of interest; this problem is considered in Appendix I.

Let us turn now to the somewhat more difficult problem of frequency modulation. It must be stated that the history of the development of the theory of modulation gives the largest number of examples of absolute confusion of concepts in connection with the incorrect application of the spectral viewpoint to oscillations. Even the comparatively simple case of amplitude modulation has served (and, for that matter, not so very long ago) as the bearer of the most erroneous expressions. Frequency modulation has fared even worse. The substance of the matter was not understood for a long time by engineers and researchers. It is instructive to reproduce here the once current idea concerning the properties of frequency modulation: in frequency modulation we have an oscillation whose frequency varies continuously within the limits of a specified interval $\omega_0 \pm \Delta\omega$; the frequency of variation of the carrier frequency depends on the modulation frequency Ω, but the amount of this variation does not. Consequently, the oscillation spectrum should be complex (since the frequency covers every value within the limits of the interval $\omega_0 \pm \Delta\omega$), while the width of the spectrum should be $2\Delta\omega$. And, since this width is arbitrarily decided by us, it is possible to shorten the band occupied by a radio-transmitting station on the frequency scale.

This is all fallacious: the spectrum obtained is not continuous, but discrete, and its width for the narrow interval $2\Delta\omega$ (the so-called swing band) does not depend entirely on the size of this interval, but is determined, as in the case of AM, by the width of the spectrum of the modulating function. Finally, when FM began to find practical application it turned out that, having considerable specific advantages, it required a band 15-20 times wider than that permitted by international law for AM.

Let us now derive the fundamental relations. We shall assume that the frequency is modulated according to the cosine law

$$\omega = \omega_0 + \Delta\omega \cos \Omega t = \omega_0 \left(1 + \frac{\Delta\omega}{\omega_0} \cos \Omega t\right),$$

where $\Delta\omega$ is the frequency decrement, while $\dfrac{\Delta\omega}{\omega_0}$ is the relative variation, i.e., the frequency modulation depth.

By its very definition the circular frequency is the time derivative of the argument of a trigonometric function representing an oscillation. Therefore we can write for the frequency-modulated oscillation, with a sinusoidal modulation,

$$x = c_0 \sin \vartheta = c_0 \sin \left(\int_0^t \omega \, dt \right) = c_0 \sin \left[\omega_0 t + \frac{\Delta\omega}{\Omega} \sin \Omega t \right] =$$

$$= c_0 [\sin \omega_0 t \cos (\beta \sin \Omega t) + \cos \omega_0 t \sin (\beta \sin \Omega t)], \tag{7.4}$$

where $\beta = \dfrac{\Delta \omega}{\Omega}$ is the so-called modulation index.

Let us consider in the beginning the relations for a small index β. Replacing the cosine and sine of the small variable in (7.4) by unity and the variable itself, respectively, we obtain

$$x \underset{\beta \ll 1}{\approx} c_0 (\sin \omega_0 t + \beta \sin \Omega t \cos \omega_0 t),$$

i.e., an expression that is in no way different from the expression for an AM oscillation. The spectrum for an FM oscillation with sinusoidal modulation, assuming a small index, just as in the case of the spectrum of an AM oscillation, consists of the carrier frequency and two satellites with frequencies $\omega_0 \pm \Omega$. Let us go on now to the general case, i.e., the case of an arbitrary index β.

Making use of the well known equations of the theory of Bessel functions[1]

$$\cos (x \sin \varphi) = J_0 (x) + 2 \sum_{n=1}^{\infty} J_{2n} (x) \cos 2n\varphi,$$

$$\sin (x \sin \varphi) = 2 \sum_{n=0}^{\infty} J_{2n+1} (x) \sin (2n + 1) \varphi,$$

$$\cos (x \cos \psi) = J_0 (x) + 2 \sum_{n=1}^{\infty} (-1)^n J_{2n} (x) \cos 2n\psi,$$

$$\sin (x \cos \psi) = 2 \sum_{n=0}^{\infty} (-1)^n J_{2n+1} (x) \cos (2n + 1) \psi,$$

we find

$$x = c_0 \Big\{ \sin \omega_0 t \, [J_0 (\beta) + 2 \sum_{n=1}^{\infty} J_{2n} (\beta) \cos 2n \, \Omega t] +$$

$$+ \cos \omega_0 t \, [2 \sum_{n=0}^{\infty} J_{2n+1} (\beta) \sin (2n + 1) \, \Omega t] \Big\}.$$

[1] See, e.g., R. O. Kuz'min, Bessel Functions [in Russian] (ONTI, 1935), p. 120. The last two equations can be obtained by assuming $\varphi = \dfrac{\pi}{2} - \psi$.

Carrying the multiplication inside the summation sign, we finally obtain

$$x = c_0 \Big\{ J_0(\beta) \sin \omega_0 t + \\ + \sum_{k=1}^{\infty} J_k(\beta) [\sin(\omega_0 + k\Omega) t + (-1)^k \sin(\omega_0 - k\Omega) t] \Big\}. \tag{7.5}$$

We thus have an oscillation with a line spectrum. Unlike AM, in this case with sinusoidal modulation an infinite spectrum arises. However, in practice, it is bounded. The essence of the matter, as we can see, is contained in the fact that the amplitudes of the harmonics are proportional to $J_k(\beta)$, and these functions have a characteristically individual property: they maintain a very small value up to some value of β which is larger the higher the order $\underline{k}$.

In practice the concept of the real bandwidth occupied by the spectrum of the FM oscillation is used. The real width is the interval on the frequency scale outside of which the harmonics have a relative magnitude ≤ 0.01. Making use of tables of Bessel functions, the limits of this interval can be found, along with its dependence on the modulation index β.

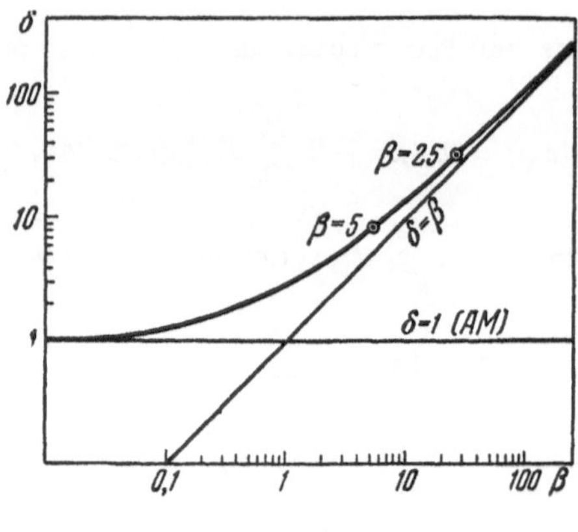

Fig. 7.

This dependence obviously has a stepwise character: its visual graph is shown in Fig. 7. The ratio, designated by δ, of half the real width of the spectrum of the modulated oscillation (i.e., the width of one side band) to the width of the spectrum of the modulating function is plotted along the vertical axis. The modulation index β is plotted on the horizontal. It must be noted that the dependence illustrated in Fig. 7 can be approximated with an accuracy sufficient for practical requirements by the simple equation

$$\delta = 1 + \beta.$$

The curve for the dependence of δ on β is drawn within the angle formed by two straight lines. The horizontal line $\delta = 1$ corresponds to the conditions appropriate for amplitude modulation; as we already know, in AM the bandwidth occupied by the modulated spectrum is always equal to twice the width of the spectrum of the modulating function. The sloped line $\delta = \beta$ is the asymptote of our curve for increasing β. As we can see, in the limit the width of the spectrum actually becomes equal to the swing band. However, this relation is strictly achieved for very large values of β, whereas in the discussion presented above it was expected that this relation could be obtained for very small values of β, and it was on this basis that it was assumed possible to contract the frequency band as the result of the application of FM. In actuality, however, in FM the quantity δ is always greater than unity; in other words, the frequency band occupied by the spectrum in connection with FM is always greater than the frequency band in connection with AM. Only for extremely small values of β do the bands become equal.

For small values of β the spectrum of the modulated oscillation in FM is rather poor, differing almost not at all from the spectrum for AM – with sinusoidal modulation it consists practically of two side lines (the rest being very small).

This situation is preserved as long as δ differs little from unity. After that the number of harmonics attaining a marked amplitude increases. The quantity δ is a direct expression of the number of these harmonics.

It is interesting to trace the form of a spectrum for large values of δ and β. The spectra for sinusoidal FM are shown in Fig. 8 for $\beta = 5$ and 25.

Considering the largest value to be $\Delta f = \frac{1}{2\pi}\Delta\omega = 75$ kc, we find that $\beta = 5$ corresponds to a modulating frequency $F = \frac{1}{2\pi}\Omega = 15$ kc, while $\beta = 25$ corresponds to a frequency $F = 3$ kc. In Fig. 8 the swing band $2\Delta\omega$ is marked off. As we can see, the real width of the spectrum, which is equal to $2\delta\Omega$, approaches the swing band. The quantity $2\delta\Omega$ is also marked off in the figure. The two spectra in Fig. 8 correspond to the relations between δ and β marked by the corresponding two points on the curve in Fig. 7.

The reader who is not accustomed to dealing with oscillations having a variable frequency will perhaps experience difficulty in the interpretation of the expression

$$\sin \vartheta = \sin\left(\int_0^t \omega\, dt\right).$$

We shall clarify this point, considering a simple example. Let us suppose that there exists a correctly recorded phonogram of a pure tone, i.e., a sinusoidal oscillation. Now let this phonogram be reproduced on an apparatus in which the linear rate of advancement of the phonogram through it is not constant. The distortion arising in this event, as we know, represents nothing more than the result of frequency modulation.

Let us derive the corresponding relations. We shall assume that the oscillation is written

$$y = \sin \omega_0 t.$$

The phonogram obtained with a constant linear velocity v_0 is expressed by the equation

$$z = \sin 2\pi \frac{x}{\lambda},$$

where $x = v_0 t$ is the coordinate calculated along the phonogram $\lambda = 2\pi \frac{v_0}{\omega_0}$ is the length of the recorded wave. Now let the phonogram move during its reproduction with a variable velocity $v(t)$. The recording mechanism, for example, an optical slit, detects the quantity

$$\zeta = \sin 2\pi \frac{\xi}{\lambda},$$

where ζ is the ordinate the phonogram at the point ξ; ξ is the path covered by the slit relative to the phonogram (Fig. 9). But this path, obviously, is equal to

$$\xi = \int_0^t v\, dt$$

and, consequently,

$$\zeta = \sin \frac{2\pi}{\lambda}\left(\int_0^t v\, dt\right).$$

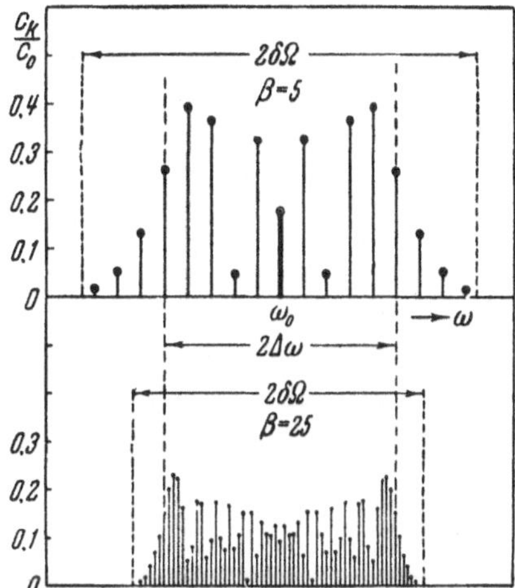

Fig. 8.

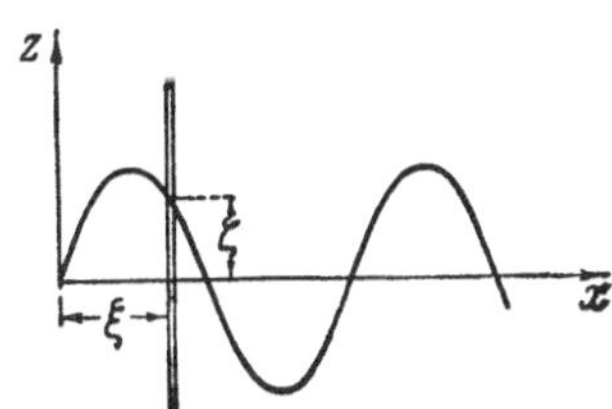

Fig. 9.

(This is perfectly analogous to the expression $\sin\vartheta = \sin\left(\int_0^t \omega\, dt\right)$. But then ϑ is that angle through which a vector turning with a variable angular velocity ω is rotated during the time t). If, for example,

$$v = v_0\left(1 + \frac{\Delta v}{v_0}\cos\Omega t\right),$$

then

$$\zeta = \sin\frac{2\pi}{\lambda}\left[\int_0^t (v_0 + \Delta v\cos\Omega t)\, dt\right] =$$

$$= \sin\frac{2\pi}{\lambda}\left(v_0 t + \frac{\Delta v}{\Omega}\sin\Omega t\right) = \sin\left(\omega_0 t + \frac{\Delta\omega}{\Omega}\sin\Omega t\right),$$

i.e., there is a typical frequency-modulated oscillation [see (7.4)]. In this expression

$$\Delta\omega = \frac{2\pi}{\lambda}\Delta v = \omega_0\frac{\Delta v}{v_0}.$$

Let us turn now to phase modulation. From the general definition we should, having taken the unmodulated oscillation

$$x = c_0 \sin(\omega_0 t + \varphi_0),$$

replace the following quantity in it:

$$\varphi = \varphi_0 + \Delta\varphi \cdot f(t).$$

Then we obtain

$$x = c_0 \sin[\omega_0 t + \varphi_0 + \Delta\varphi \cdot f(t)]. \tag{7.6}$$

For frequency modulation, introducing the variable frequency

$$\dot{\omega} = \omega_0 + \Delta\omega \cdot f(t),$$

we would obtain

$$\begin{aligned} x &= c_0 \sin\left[\omega_0 t + \varphi_0 + \Delta\omega \int_0^t f(t)\,dt\right] = \\ &= c_0 \sin[\omega_0 t + \varphi_0 + \Delta\omega \cdot F(t)]. \end{aligned} \tag{7.7}$$

Comparing (7.6) and (7.7), we see that FM and PM in essence give a perfectly identical form of oscillation. The only difference is contained in the fact that in the case of PM in the argument of the sinusoidal function there appears the modulating function $f(t)$, while in FM its integral appears.

If the modulation is sinusoidal, then differences in the form of the modulated oscillations and their spectra are in general considered impossible, since the integral of a sinusoid is a cosinusoid, i.e., another sinusoid, but one that is shifted in phase by the amount $\pi/2$. There is nevertheless a difference, but a more subtle one. Its essence is contained in the fact that if

$$f(t) = \sin\Omega t,$$

then

$$F(t) = -\frac{1}{\Omega}\cos\Omega t.$$

The appearance of the factor $\frac{1}{\Omega}$ indicates the necessity for introducing appropriate correction factors.

If we assume that the discriminators, i.e., elements that convert PM and FM into AM, run the same way, then a single integrating link must be added to the later circuit of the PM receiver in comparision with the same circuit for the FM receiver.

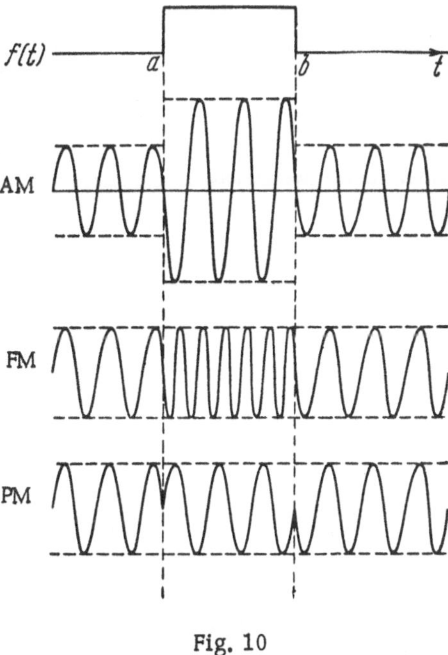

Fig. 10

Further comparison of PM and FM would carry us very deep into specialized areas of modern electronic engineering. We shall limit our discussion to general considerations and only attempt in conclusion to explain the difference between all three forms of modulation in as easily visualized a manner as possible.

As already stated, in sinusoidal FM and PM a difference in the form of the modulated oscillations is considered impossible. But this difference does arise when the modulation becomes very complex, i.e., when the modulating function has a more or less rich spectrum.

Let us assume that the modulating function is a pulse in the following form:

$$f(t) = \begin{cases} 0 & (t < a), \\ 1 & (a < t < b), \\ 0 & (t > b). \end{cases}$$

In this case, when t = a the amplitude in AM, the frequency in FM, and the phase in PM are changed by a sudden jump.

A change by means of a sudden jump back to the original values takes places for t = b. This is illustrated in Fig. 10. As we can see, there exists a difference between the modulated oscillations in the cases of FM and PM. Along the same line, we note that a discontinuous change in phase can be interpreted as the result of an infinitely rapid change in frequency over the extent of an infinitesimal time interval. In the last analysis all this reduces once again to the fundamental definitions

$$\omega = \frac{d\vartheta}{dt};$$

$$\vartheta = \int \omega \, dt.$$

8. Spectrum Transfer

In technology it is of great interest to examine the possibility of a transformation of a specific function of time in such a way that the spectrum of the function would be displaced along the frequency scale. During ordinary modulation or frequency conversion two side bands, as we know, are obtained. The question is asked, how can a spectrum consisting of only one side band be obtained? The required spectral transformation is shown in Fig. 11. We assume that the initial spectrum is bounded above by the frequency ω_c.

It must be explained from the outset that the displacement theorem (Section 4) does not give the required transformation, since it refers to a shift in a complex spectrum $S(\omega)$, whereas we are required to displace a real amplitude spectrum $\Phi(\omega) = |S(\omega)|$ (see [9]).

The simplest method for obtaining a displaced spectrum is to first form a two-band modulation spectrum by means of balanced modulation of the carrier frequency, then to suppress the lower side band by means of filters. There are other possibilities of interest, two of which we shall consider presently.

We shall proceed from Eq. (2.14):

$$f(t) = \frac{1}{\pi} \operatorname{Re} \int_0^\infty S(\omega)\, e^{j\omega t}\, d\omega. \tag{8.1}$$

It is necessary to construct some function that will have the spectral density $S(\omega)$ at the frequency $\omega_0 + \omega$. For such a function we can write

$$f_1(t) = \frac{1}{\pi} \operatorname{Re} \int_0^\infty S(\omega)\, e^{j(\omega_0+\omega)t}\, d\omega. \tag{8.2}$$

This expression can be transformed as follows:

$$f_1(t) = \frac{1}{\pi} \operatorname{Re} e^{j\omega_0 t} \int_0^\infty S(\omega)\, e^{j\omega t}\, d\omega =$$

$$= \frac{1}{\pi}\left(\cos \omega_0 t \operatorname{Re} \int_0^\infty S(\omega)\, e^{j\omega t}\, d\omega - \sin \omega_0 t \operatorname{Im} \int_0^\infty S(\omega)\, e^{j\omega t}\, d\omega\right)$$

or

$$f_1(t) = f(t) \cos \omega_0 t + f^{\vee}(t) \sin \omega_0 t^{1}. \tag{8.3}$$

Let us define the function

$$f^{\vee}(t) = -\frac{1}{\pi} \operatorname{Im} \int_0^\infty S(\omega)\, e^{j\omega t}\, d\omega. \tag{8.4}$$

[1]The complex function

$$F(t) = \frac{1}{\pi} \int_0^\infty S(\omega)\, e^{j\omega t}\, d\omega,$$

whose real and imaginary components are of concern here, is called the "analytic signal" and proves to be quite useful in certain theoretical investigations.

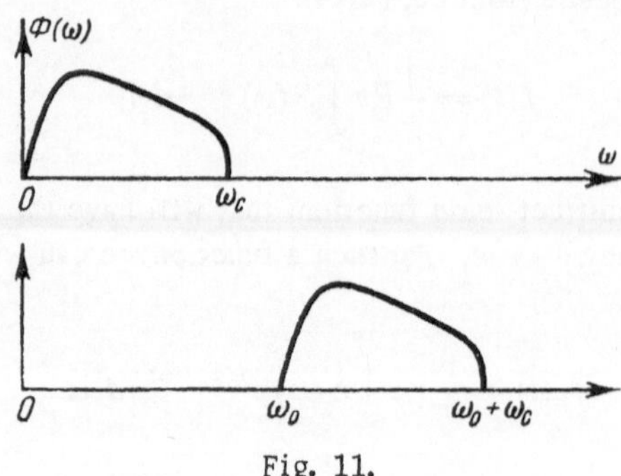

Fig. 11.

Then let us write the complex spectrum, for the present purpose, in the form

$$S(\omega)=A(\omega)+jB(\omega).$$

Substituting into (8.4), we obtain

$$f^{\vee}(t)=-\frac{1}{\pi}\int_0^{\infty}(A\sin\omega t+B\cos\omega t)\,d\omega= \tag{8.5}$$

$$=\frac{1}{\pi}\int_0^{\infty}\left[A\cos\left(\omega t+\frac{\pi}{2}\right)-B\sin\left(\omega t+\frac{\pi}{2}\right)\right]d\omega,$$

while for the function f(t) we obtain by an analogous approach

$$f(t)=\frac{1}{\pi}\int_0^{\infty}(A\cos\omega t-B\sin\omega t)\,d\omega. \tag{8.6}$$

Comparing (8.5) and (8.6), we see that the function $f^{\vee}(t)$ differs from the function f(t) in that all the components are displaced in phase by $\pi/2$. Consequently, for the formation of the function $f^{\vee}(t)$, it is necessary to impose a phase inverter, which shifts the phase by $\pi/2$ for all frequencies, i.e., a quadripole whose complex transmission coefficient is equal to

$$K=j=e^{j\frac{\pi}{2}}. \tag{8.7}$$

A quadripole with such properties cannot be physically realized; it is possible, however, to construct a real quadripole in which the characteristic (8.7) is brought about with any desired approximation in a finite frequency interval (except for $\omega = 0$ and the neighboring region of this value).

The required transformation (8.3) is fulfilled by the selective circuit shown in Fig. 12, in which: P is a $\pi/2$ phase inverter; G is a two-phase generator giving two oscillations displaced in phase by $\pi/2$: $\sin \omega_0 t$ and $\cos \omega_0 t$; M_1 and M_2 are multipliers (balanced modulators).

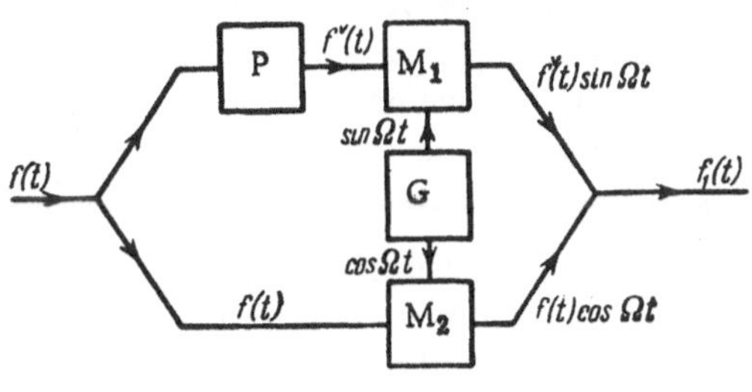

Fig. 12.

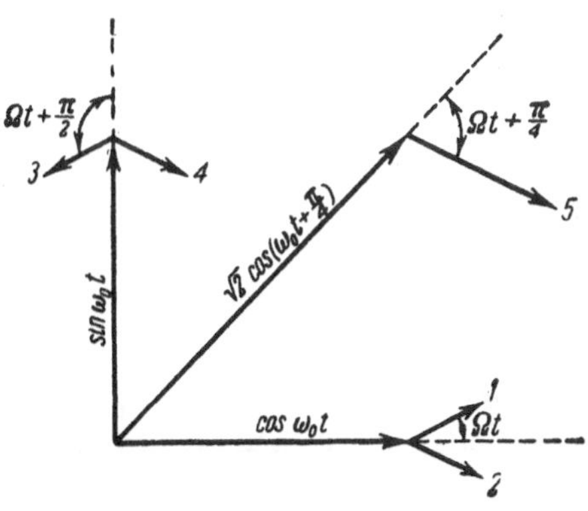

Fig. 13.

The essential part of the transformation (8.3) appears as the vector diagram shown in Fig. 13, in which the following case is shown:

$$f(t) = \cos \Omega t, \quad f^{\vee}(t) = -\sin \Omega t = \cos\left(\Omega t + \frac{\pi}{2}\right).$$

For the purpose of visualization we have not assumed a balanced modulation, but instead ordinary AM, so that the ω_0 component of the carrier frequency is not eliminated. The vector of the carrier oscillation $\cos\omega_0 t$ is plotted on the horizontal; as a result of modulation by the function f(t) there appear two satellites, represented by the vectors 1 and 2, which are situated at an angle $\pm\Omega t$ relative to the fundamental vector. The vector $\sin\omega_0 t$ is plotted on the vertical; modulation by the function $f^{\vee}(t)$ gives two satellites, the vectors of which, 3 and 4, are situated at an angle $\pm\left(\Omega t+\frac{\pi}{2}\right)$ relative to the fundamental vector. The directions of the vectors 2 and 4 coincide, while the directions of the vectors 1 and 3 are opposite to one another. As the result of addition the carrier-frequency vector and one satellite, the vector 5, are obtained. The second satellite is eliminated, and this, then, is single-band sinusoidal modulation.[1]

It can be shown that the consideration of this transformation is a special case of multiphase modulation, so that the described mechanism can be called two-phase single-band modulation. From the engineering point of view three-phase single-band modulation is of very great interest [17].

Let us now consider another possibility for transferring a spectrum. A spectrum that is bounded above by the frequency ω_c can be represented by a Fourier series (see Section 14):

$$S(\omega)=\Delta t\sum_{-\infty}^{\infty} f_k e^{-jk\omega\Delta t}, \tag{8.8}$$

where

$$\Delta t=\frac{\pi}{\omega_c}, \quad f_k=f(k\,\Delta t).$$

The spectrum $S_1(\omega)$, which is shifted by the amount ω_0 along the frequency scale, is connected with the original spectrum $S(\omega)$ by the relation

$$S_1(\omega_0+\omega)=S(\omega),$$

so that

$$S_1(\omega)=\Delta t\sum_{-\infty}^{\infty} f_k e^{jk(\omega_0-\omega)\Delta t}. \tag{8.9}$$

[1]The vector diagram shows that in the presence of a carrier, single-band modulation can be considered as complex amplitude-phase modulation, serving as a basis for obtaining single-band modulation by one of the methods that we are not considering.

The time function corresponding to this spectrum is expressed by the relation

$$f_1(t) = \frac{1}{\pi} \operatorname{Re} \int_{\omega_0}^{\omega_0+\omega_c} S_1(\omega)\, e^{j\omega t}\, d\omega . \tag{8.10}$$

Substituting here the expression (8.8) for the spectrum, we find [8, 9]

$$f_1(t) = \sum_{-\infty}^{\infty} f_k \frac{\sin \frac{\omega_c}{2}(t - k\,\Delta t)}{\frac{\omega_c}{2}(t - k\,\Delta t)} \cos\left(\omega_1 t - k\,\frac{\pi}{2}\right), \tag{8.11}$$

where

$$\omega_1 = \omega_0 + \frac{1}{2}\,\omega_c$$

is the center frequency of the transfered spectrum. For the original function f(t), [whose spectrum is expressed by Eq. (8.8)], we have

$$f(t) = \sum_{-\infty}^{\infty} f_k \frac{\sin \omega_c (t - k\Delta t)}{\omega_c (t - k\,\Delta t)} . \tag{8.12}$$

Consideration of Eq. (8.11) indicates the nature of the transformation necessary for obtaining the function $f_1(t)$ with a displaced spectrum from the original function f(t). Such a transformation can be carried out by the spectral circuit shown in Fig. 14. In this circuit we have: PG is a pulse generator with a pulse-repetition period Δt; RA is a recording apparatus, the purpose of which is to take readings of the multiple values of the function f(t) at the instants that pulses are delivered from the PG, i.e., to determine the magnitude of f_k and to deliver a pulse with an included area proportional to f_k. Moreover, it is required that these pulses, first of all, have the required sign and, secondly, that they be delivered alternately to the two outputs of the apparatus, i.e., that the even-numbered pulses, for example, be delivered to the upper output, while the odd-numbered ones are delivered to the lower. Then follow the low-frequency filters LFF with the upper bounding frequency $\frac{1}{2}\omega_c$. After that come two multipliers M, to the secondary inputs of which potentials displaced in phase by $\pi/2$ and having the frequency $\omega_1 = \omega_0 + \frac{1}{2}\omega_c$ are delivered from the two-phase generator G. The potentials taken from both multipliers are compounded.

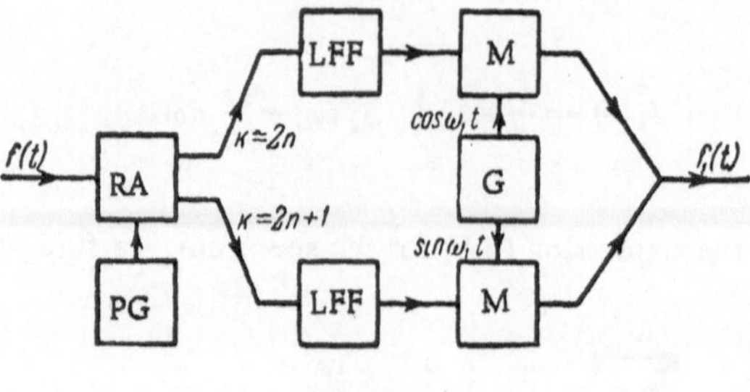

Fig. 14.

The circuit described is of interest in that there is no wideband phase inverter in it. However, in order to obtain good results it is obviously necessary to impose rigorous requirements on the low-frequency filters.

9. Transformation of Spectra in Detection

If some periodic oscillation x is subjected to the nonlinear operation

$$y = \varphi(x),$$

then the oscillation y obtained as a result of this operation will have a spectrum that differs from the spectrum of x and, as a rule will be richer. Thus, for example, if the original oscillation is the sum of two sinusoids

$$x = c_1 \sin \omega_1 t + c_2 \sin \omega_2 t$$

and, as a consequence, has a spectrum consisting of two spectral lines, then after the nonlinear operation φ, we obtain in the composition of the oscillation y spectral components with the frequencies

$$\omega_{mn} = m\omega_1 \pm n\omega_2,$$

where m and n in the general case are any positive integers. This type of spectrum is called a combination (Raman) spectrum, and the frequencies ω_{mn} are called the combination frequencies.

Such a modification of a spectrum is used in measuring the degree of deviation of a given system from linearity. The oscillation x is delivered to the input of the system; the oscillation y obtained at the output is then studied. Combination frequencies do not arise in the singular case when φ (which, in a given case, expresses the characteristic of the investigated system) is a linear function.

A special form of nonlinear operation used to transform a spectrum is called the detection operation. In the composition of a modulated oscillation there is no component with the modulation frequency. But this component is not necessary for our purposes, since it is nothing more than the transmitted signal. In order to make it show up again it is necessary to subject the modulated oscillation to some nonlinear operation. This operation, which serves the purpose of forming a component with the modulation frequency, is called detection. As the result of detection a complex oscillation is obtained, in which the oscillation of interest here, with the modulation frequency, appears as one of the components. Further separation of the components does not present any particular difficulty.

A consideration of the problems of detection in general form would be exceedingly cumbersome; we shall limit ourselves to some simpler examples.

Let us consider a simple AM oscillation with sinusoidal modulation. We have orginally an oscillation with the carrier frequency $\sin \omega_0 t$ and the modulating oscillation $1 + m \sin \Omega t$. As the result of some operation, which we call modulation, both of these oscillations turn out to be multiplied together, and we have

$$x = (1 + m \sin \Omega t) \sin \omega_0 t.$$

In the composition of the modulated oscillation, as we know, there is no longer a component with the frequency Ω; the spectrum of x consists of three lines with the frequencies ω_0, $\omega_0 + \Omega$ and $\omega_0 - \Omega$.

If we now desire to obtain once again the oscillation with frequency Ω we must apply the appropriate detection to x. The detection operation in the given case performs an action which is inverse to the operation of modulation, so that in application to a modulated signal detection is sometimes called demodulation. We shall consider only two fundamental forms of detector: the "linear" detector

$$y = |x|$$

(Fig. 15a), and the square-law detector

$$y = x^2,$$

(Fig. 15b). The word linear is given in quotation marks above, in order to emphasize the fact that the linear detector is actually nonlinear and that a linear detector in the true sense is impossible (i.e., a linear system is incapable of detection).

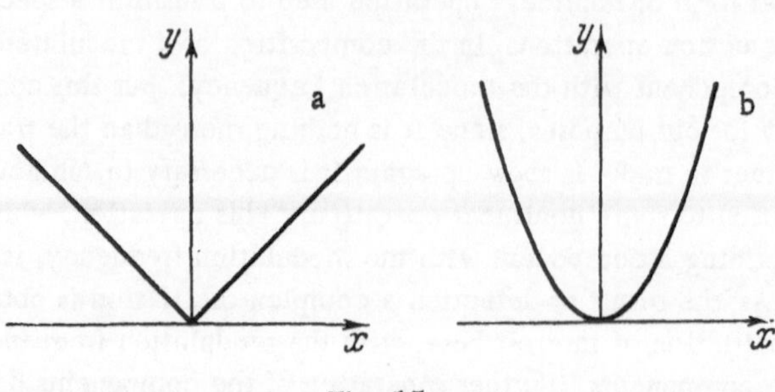

Fig. 15.

For the detection of a modulated oscillation it is convenient to use linear detection. Applying the fact that the absolute value of a product is equal to the product of the absolute values, we can write

$$y = |x| = (1 + m \sin \Omega t)\,|\sin \omega_0 t|.$$

However, the function $|\sin \omega_0 t|$ can be represented by the following Fourier series:

$$|\sin \omega_0 t| = \frac{2}{\pi}\left(1 - \sum_{k=1}^{\infty} \frac{1}{4k^2 - 1} \cos 2k\omega_0 t\right),$$

whence

$$y = |x| = \frac{2}{\pi}\left\{(1 + m \sin \Omega t) - \sum_{k=1}^{\infty} \frac{1}{4k^2 - 1}\left[\cos 2k\omega_0 t + \right.\right.$$

$$\left.\left. + \frac{m}{2} \sin\,(2k\omega_0 + \Omega)\,t - \frac{m}{2} \sin\,(2k\omega_0 - \Omega)\,t\right]\right\}.$$

In this expression the first term (in the parentheses) is the modulating function, which we are striving to obtain; the second term joins the higher-frequency components $2k\omega_0$, $2k\omega_0 + \Omega$ and $2k\omega_0 - \Omega$, which are easily separated, under the summation sign.

If we were to subject the modulated oscillation to square-law detection, we would obtain

$$y = x^2 = (1 + m \sin \Omega t)^2 \sin^2 \omega_0 t =$$
$$= \frac{1}{2}\left\{1 + \frac{m^2}{2} + 2m \sin \Omega t - \frac{m^2}{2} \cos 2\Omega t - \left(1 + \frac{m^2}{2}\right) \cos 2\omega_0 t - \right.$$
$$- m \sin (2\omega_0 - \Omega)\, t + m \sin (2\omega_0 + \Omega)\, t +$$
$$\left. + m^2 \cos 2\, (\omega_0 - \Omega)\, t + m^2 \cos 2\, (\omega_0 + \Omega)\, t \right\}.$$

Thus in this case, in addition to the constant component $1 + \frac{m^2}{2}$ and the five spectral lines with the high frequencies

$$2\omega_0,\ 2\omega_0 - \Omega,\ 2\omega_0 + \Omega,\ 2\,(\omega_0 - \Omega),\ 2\,(\omega_0 + \Omega),$$

we obtain two spectral lines with the low frequencies Ω and 2Ω. Consequently, the spectrum of the modulating oscillation, which is composed of the single original line with frequency Ω, turns out to be distorted, and this form of detection can be applied only with a very small modulation depth (since the amplitude ratio of the second and first harmonics is equal to $m/4$).

The cases dissected here are shown in Fig. 16. In Fig. 16a the two original oscillations—with the carrier frequency ω_0 and modulating frequency Ω – are shown in Fig. 16b, the spectrum of the modulated oscillation—the carrier frequency and two satellites—is shown; in Fig. 16c the spectrum of the oscillation obtained as a result of linear detection of the modulated oscillation is shown (the appearance of the line with frequency Ω must be noted). Finally, in Fig. 16d the spectrum obtained as a result of square-law detection is shown (the spectrum is bounded, but has two low-frequency lines at Ω and 2Ω).

Let us consider the problem of detecting beats. "Beats" refer to an interference effect consisting of the periodic variation of the amplitude of an oscillation composed of two simple sinusoidal oscillations with unequal frequencies. It is usually stated that the beat frequency is equal to the difference in frequencies of the generating oscillations.[1]

[1]This is true when the beat frequency is understood to mean the frequency of repetition of maxima or minima in the envelope of the resultant oscillation. This is not true when the beat frequency is understood to mean the fundamental frequency of the observed process. When the frequencies of the generating oscillations are incommensurate the resultant oscillation is, in general, nonperiodic.

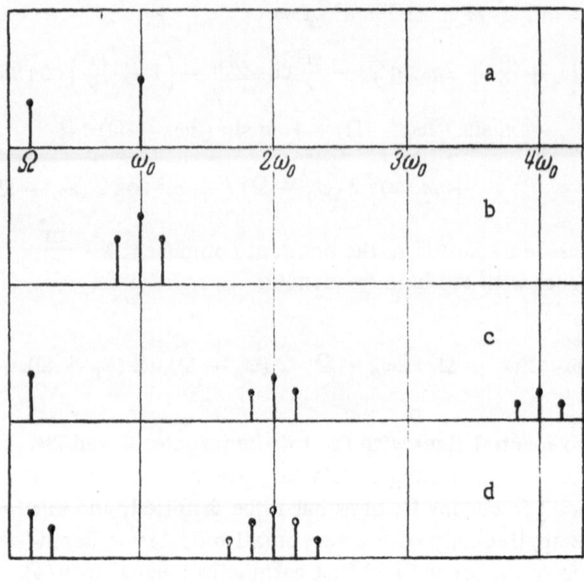

Fig. 16.

Let us assume that we have before us the problem of obtaining a sinusoidal oscillation with a different frequency as the result of detection. We have

$$x = \sin \omega_1 t + \sin \omega_2 t.$$

In this case square-law detection must be applied, giving

$$y = x^2 = \sin^2 \omega_1 t + \sin^2 \omega_2 t + 2 \sin \omega_1 t \sin \omega_2 t =$$
$$= 1 - \frac{1}{2} \cos 2\omega_1 t - \frac{1}{2} \cos 2\omega_2 t - \frac{1}{2} \cos (\omega_1 + \omega_2) t +$$
$$+ \frac{1}{2} \cos (\omega_1 - \omega_2) t.$$

As we can see, we have obtained, in addition to the constant component and high frequencies $2\omega_1$, $2\omega_2$ and $\omega_1 + \omega_2$, the required oscillation with frequency $\omega_1 - \omega_2$. This cannot result from linear detection. Applying the latter, we would obtain

$$y = |x| = |\sin \omega_1 t + \sin \omega_2 t| =$$
$$= 2 \left| \sin \frac{\omega_1 + \omega_2}{2} t \right| \left| \cos \frac{\omega_1 - \omega_2}{2} t \right| =$$
$$= \frac{8}{\pi^2} \left[1 - \sum_{k=1}^{\infty} \frac{1}{4k^2 - 1} \cos k (\omega_1 + \omega_2) t \right] \times$$
$$\times \left[1 - \sum_{k=1}^{\infty} \frac{(-1)^k}{4k^2 - 1} \cos k (\omega_1 - \omega_2) t \right]$$

i.e., in addition to the constant component and high frequencies, we would have an infinite spectrum of low-frequency components with frequencies $k(\omega_1 - \omega_2)$.

Square-low detection in connection with beat detection gives the necessary result also in the more complex case when the amplitudes of the generating oscillations are not equal, i.e., when

$$x = \sin \omega_1 t + \varepsilon \sin \omega_2 t$$

($\epsilon \gtrless 1$). The expression for the beat envelope when the amplitudes are equal is

$$c(t) = \left| \cos \frac{\omega_1 - \omega_2}{2} t \right|$$

If the amplitudes are not equal, then for the envelope we obtain

$$c(t) = \sqrt{1 + \varepsilon^2 + 2\varepsilon \cos (\omega_1 - \omega_2) t}$$

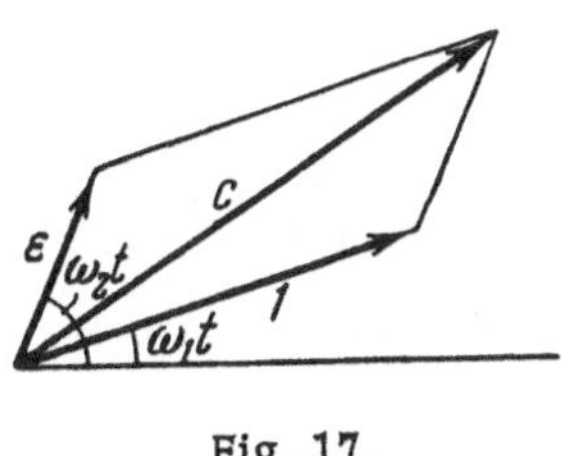

Fig. 17.

This expression is most simply obtained from a geometric construction (Fig. 17). Only at very small limits or at very large values of ϵ does the beat envelope approach a sinusoid. When $\epsilon \ll 1$

$$c(t) = \sqrt{1 + \varepsilon^2 + 2\varepsilon \cos (\omega_1 - \omega_2) t} \approx$$
$$\approx \sqrt{1 + 2\varepsilon \cos (\omega_1 - \omega_2) t} \approx$$
$$\approx 1 + \varepsilon \cos (\omega_1 - \omega_2) t,$$

and, consequently, under such conditions linear detection can also be applied.

The beat curve in this case is no different from the curve for a sinusoidally modulated oscillation when the modulation depth is small. Exactly the same situation exists when $\epsilon \gg 1$.

10. Spectrum of a Sum of Periodic Functions

In Section 4 the fact was already mentioned that the Fourier transform is linear, so that the principle of superposition can be applied to it. In the case of periodic functions this can be written in the following way:

$$C_k = \frac{1}{T} \int_{-\frac{T}{2}}^{+\frac{T}{2}} dt \sum f_i(t)\, e^{-j2\pi k \frac{t}{T}} = \sum_i C_{ik},$$

i. e., the complex amplitude of the kth harmonic of the spectrum of a summation of functions is equal to the sum of the kth harmonics of the spectra of each separate function. This is without contradiction, but we are usually mostly concerned with real amplitudes. For these we can write

$$c_k = 2\,|C_k| = 2\,|\sum_i C_{ik}|\,.$$

From the geometric point of view this quantity represents the closing of a polygon, the sides of which are equal to c_{ik} and are oriented at the corresponding angles φ_{ik}. Let us assume that two sinusoidal oscillations are specified with the complex amplitudes

$$2C_1 = c_1 e^{-j\varphi_1}, \quad 2C_2 = c_2 e^{-j\varphi_2}\,.$$

Then

$$2C = 2(C_1 + C_2) = c_1 e^{-j\varphi_1} + c_2 e^{-j\varphi_2}\,,$$

$$\begin{aligned} c &= |c_1 e^{-j\varphi_1} + c_2 e^{-j\varphi_2}| = \\ &= |c_1 \cos\varphi_1 + c_2 \cos\varphi_2 - j(c_1 \sin\varphi_1 + c_2 \sin\varphi_2)| = \\ &= \sqrt{(c_1\cos\varphi_1 + c_2\cos\varphi_2)^2 + (c_1\sin\varphi_1 + c_2\sin\varphi_2)^2} = \\ &= \sqrt{c_1^2 + c_2^2 + 2c_1c_2\cos(\varphi_1 - \varphi_2)}. \end{aligned} \tag{10.1}$$

This equation can be applied in that case when, instead of constant phase angles φ, we have arbitrarily time-dependent angular arguments ϑ. Then the amplitude turns out to be a function of time and Eq. (10.1) acquires the sense of an envelope of some complex oscillation. For example, if in place of

φ_1 and φ_2 we substitute $\omega_1 t$ and $\omega_2 t$, then we obtain an expression for the envelope of the beats arising with the addition of two sinusoidal oscillations with amplitudes c_1 and c_2 and frequencies ω_1 and ω_2.

Let us consider the problem of the spectrum of a function obtained as the result of the superposition of two identical periodic functions which are, however, displaced in time. For some periodic function f(t) we have

$$C_{1k}=\frac{1}{T}\int_{-\frac{T}{2}}^{+\frac{T}{2}} f(t)\, e^{-j2\pi k\frac{t}{T}}\, dt.$$

For the same function retarded in time by the amount τ we have

$$C_{2k}=\frac{1}{T}\int_{-\frac{T}{2}}^{+\frac{T}{2}} f(t-\tau)\, e^{-j2\pi k\frac{t}{T}}\, dt$$

or, replacing $t-\tau$ by t_1,

$$C_{2k}=\frac{1}{T}\int_{-\frac{T}{2}}^{+\frac{T}{2}} f(t_1)\, e^{-j2\pi k\frac{t_1+\tau}{T}}\, dt_1 = e^{-jk\Omega\tau}\, C_{1k}.$$

If now we combine the functions $f(t)$ and $f(t-\tau)$, the complex amplitude of the kth harmonic of their sum will be equal to

$$C_k=C_{1k}+C_{2k}=C_{1k}\left(1+e^{-jk\Omega\tau}\right),$$

and the real amplitude will be equal to[1]

$$c_k=2\left|C_k\right|=c_{1k}\left|1+e^{-jk\Omega\tau}\right|=2c_{1k}\left|\cos\frac{k\Omega\tau}{2}\right|. \qquad (10.2)$$

[1]This equation could be obtained from (10.1) by assuming

$$c_1=c_2=c_{1k},\quad \varphi_1=k\Omega\tau,\quad \varphi_2=0.$$

Thus, in order to obtain the spectrum of a sum of two identical functions displaced in time by the amount τ (for example, the sum of a signal and its reflection), it is sufficient to multiply the amplitude of each harmonic by $2\left|\cos\frac{k\Omega\tau}{2}\right|$.

Let us take an example. Let a periodic sequence of short pulses be given and let $\tau = \frac{T}{2}$. Then the multiplier in Eq. (10.2) assumes the form

$$2\left|\cos k\frac{\pi}{2}\right| = \begin{cases} 0 & k \text{ odd} \\ 2 & k \text{ even} \end{cases}$$

Therefore, all odd harmonics drop out. This is just as it should be: if $\tau = \frac{T}{2}$, this means that the pulses of the second series fall in the middle of the intervals of the first, i.e., pulses recur twice as frequently; consequently the fundamental frequency and with it all the harmonics are doubled.

It is well understood that the same thing is obtained when $\tau = \frac{2n+1}{2}T$, i.e., when τ is equal to any odd number of half periods. If $\tau = \frac{2n+1}{4}T$, then the second, sixth, tenth, etc., harmonic drops out of the spectrum.

Equation (10.2) gives the value of the amplitude of the kth harmonic of the spectrum of a summation of the functions $f(t)$ and $f(t-\tau)$. If now we compose the difference between these two functions rather than the summation, then, proceeding in a manner similar to the preceding, we find

$$c_k = c_{1k}\left|1 - e^{-jk\Omega\tau}\right| = 2c_{1k}\left|\sin\frac{k\Omega\tau}{2}\right|. \tag{10.3}$$

Let us now assume that τ is a sufficiently small quantity that the following approximation is valid:

$$f(t) - f(t-\tau) = \Delta f(t) \approx \tau f'(t) = \overset{*}{f}(t).$$

Thus we have expressed the difference in functions in terms of a derivative. Let us determine the spectrum of $\overset{*}{f}$:

$$\overset{*}{C}_k = \frac{1}{T}\int_{-\frac{T}{2}}^{+\frac{T}{2}} \overset{*}{f}(t)\, e^{-jk\Omega t}\, dt = \frac{\tau}{T}\int_{-\frac{T}{2}}^{+\frac{T}{2}} f'(t) e^{-jk\Omega t}\, d\tau = \tau\left[\frac{f(t)}{T} e^{-jk\Omega t}\Bigg|_{-\frac{T}{2}}^{+\frac{T}{2}} + jk\Omega C_k\right].$$

But, since the function f(t) is periodic,

$$\overset{*}{C}_k = jk\Omega\tau C_k$$

and, consequently,

$$\overset{*}{c}_k = k\Omega\tau c_k.$$

This relation could be obtained from (10.3), replacing the sine by its argument.

All of the relations given above can be extended without difficulty to the case of a quasi-periodic function; in this case the quantity Ω_k appears in all the equations in place of $k\Omega$.

11. Spectra of Special Pulses

Let us first consider the spectra of some special discontinuous functions, in particular, the so-called unit function $\sigma(t)$, which is defined as follows:

$$\sigma(t) = \begin{cases} 0 & t < 0, \\ \frac{1}{2} & t = 0, \\ 1 & t > 0. \end{cases} \tag{11.1}$$

A more general definition is contained in the unit function delayed by the amount τ:

$$\sigma(t-\tau) = \begin{cases} 0 & t < \tau, \\ \frac{1}{2} & t = \tau, \\ 1 & t > \tau. \end{cases}$$

Further, the definition for a unit pulse is introduced in the form

$$\int_{-\infty}^{t} \delta(u)\, du = \sigma(t), \tag{11.2}$$

where the function $\delta(t)$ is equal to zero everywhere except at the point $t = 0$, where it is equal to infinity. Thus $\delta(t)$ is an infinitely short pulse with an area equal to unity. The function $\delta(t)$ was first introduced into the mathematical apparatus of theoretical physics by Dirac; it is also called the Dirac function, or delta function.

Differentiating (11.2) with respect to t, we obtain the formal equation[1]

$$\delta(t) = \frac{d}{dt}\sigma(t). \tag{11.3}$$

A qualitative idea of the properties of the functions σ and δ and some basis for the relation (11.3) can be gained from limiting transitions. Let us consider the auxiliary function

$$\sigma_a(t) = \begin{cases} 0 & t < -\frac{a}{2}, \\ \frac{1}{2} + \frac{t}{a} & -\frac{a}{2} < t < \frac{a}{2}, \\ 1 & t > \frac{a}{2}, \end{cases}$$

and its derivative

$$\delta_a(t) = \begin{cases} 0 & t < -\frac{a}{2}, \\ \frac{1}{a} & -\frac{a}{2} < t < \frac{a}{2}, \\ 0 & t > \frac{a}{2}. \end{cases}$$

Now we pass to the limit as $a \to 0$. We note that the area of the rectangular pulse $\delta_a(t)$ remains equal to unity, independently from the value of a (Fig. 18). In the limit we have

$$\sigma_a(t) \underset{a\to 0}{\to} \sigma(t), \qquad \delta_a(t) \underset{a\to 0}{\to} \delta(t).$$

Having settled on the basic definitions, let us proceed to the spectra. For the spectrum of the δ-function we obtain

---

[1]The operation of differentiation, as performed here on the discontinuous function σ, is considered legitimate nowadays in mathematics, in connection with the introduction of so-called generalized functions.

$$S_1 = \int_{-\infty}^{+\infty} e^{-j\omega t}\delta(t-\tau)\,dt = e^{-j\omega\tau}\ .$$

Thus the absolute value of the spectrum of the δ-function is equal to unity. This means that the δ-function possesses a complex spectrum extending to infinitely large values of the frequency with an invariant spectral density.

In attempting to compute the spectrum of the function σ, difficulties arise in that this function is not absolutely integrable and has a finite value at infinity. This difficulty can be circumvented by multiplying the function σ by $e^{-\alpha t}$.[1]

[1]Sometimes the following is simply written:

$$\int_0^{\infty} e^{-j\omega t}\,dt = \frac{1}{j\omega},$$

i.e., it is assumed that $e^{-j\infty} = 0$, which then requires a certain reservation.

When we write $f(\infty)$ it is understood that we mean

$$f(\infty) = \lim_{x\to\infty} f(x).$$

The function e^{-jx} when x tends to infinity does not tend to zero, nor does it, in general, tend to any definite limit at all. Therefore, the expression $e^{-j\infty}$ is devoid of any meaning. The function e^{-jx} (which, in other respects, is like e^{+jx}) has an absolute value equal to unity for any value of x, the argument varying cyclically with increasing x. Another matter comes up when the exponent is a complex quantity, i.e., if, for example, the function is written in the form

$$f(x) = e^{(-\alpha+j\beta)x}.$$

Then when α is arbitrarily small, but finite, real, and positive, we can write, in the sense indicated above,

$$f(\infty) = 0.$$

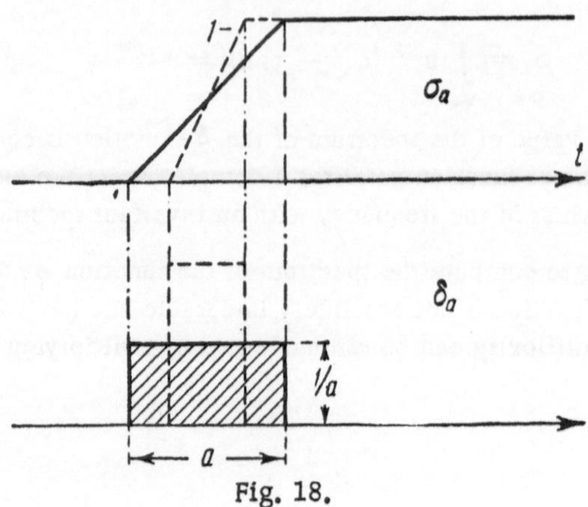

Fig. 18.

Then

$$S_0 = \int_{-\infty}^{+\infty} e^{-(\alpha+j\omega)t}\sigma(t-\tau)\,dt = \int_{\tau}^{\infty} e^{-(\alpha+j\omega)t}\,dt =$$

$$= \frac{1}{\alpha+j\omega}e^{-(\alpha+j\omega)\tau} \underset{\alpha\to 0}{\to} \frac{1}{j\omega}e^{-j\omega\tau}$$

The modulus of the spectrum of the function σ is $\frac{1}{\omega}$. It is possible to arrive at this same result in another way. Let us apply the well known relation.[1]

$$\sigma(t) = \frac{1}{2} + \frac{1}{\pi}\int_{0}^{\infty}\frac{\sin\omega t}{\omega}\,d\omega.$$

The integral on the right-hand side can be considered the real form of the Fourier integral for the odd function $\sigma(t) - \frac{1}{2}$. In this case the spectrum of this function (i.e., the multiplier associated with the sine term) is $\frac{1}{\omega}$.

Let us now consider the spectrum of a very short pulse of arbitrary form. This pulse is characterized by the fact that the function that it represents is

[1]See, e.g., V. I. Smirnov, Course in Higher Mathematics [in Russian] (Gostekhizdat, 1948) (ninth edition), Vol. 2, pp. 252-253.

equal to zero outside some very small inverval τ about $t = 0$.[1]

For the spectrum of such a pulse we can write

$$S = \int_{-\infty}^{+\infty} e^{-j\omega t} f(t)\,dt = \int_{-\frac{\tau}{2}}^{+\frac{\tau}{2}} e^{-j\omega t} f(t)\,dt.$$

But if τ is small $e^{\pm j\omega\frac{\tau}{2}}$ is not much different from unity and

$$S \approx \int_{-\frac{\tau}{2}}^{+\frac{\tau}{2}} f(t)\,dt = q,$$

i.e., the spectrum is equal to a constant, defined by the area of the pulse. We have already encountered such a relation in developing the problem of the spectrum of the unit pulse δ (t). However, in that case we were concerned with some known, definite function, whereas here the function can have an arbitrary form in the interval wherein it exists, i.e., the pulse can have any form whatsoever, only that its duration must be small.

In this case this problem plays a basic role, and we shall dwell on it somewhat further. We shall demand that τ be small. However, such a requirement is void of meaning until some quantity is specified for comparison. In the present instance, the criterion is determined by the fact that when τ is small the function $e^{\pm j\omega\frac{\tau}{2}}$ differs little from unity. But this is satisfied by the condition

$$\omega\frac{\tau}{2} \ll 1$$

or

$$\tau \ll T,$$

where T is the period corresponding to the frequency ω $\left(T = \frac{2\pi}{\omega}\right)$.

[1] It is entirely optional that the pulse is assumed to be at the origin of the time scale: as we know, the spectrum does not depend on any shift in time.

Thus we have arrived at a conclusion which is very simple, but which is very important, both from the hypothetical and from the practical point of view: a single pulse with arbitrary form has a complex spectrum that can be approximately expressed by a constant which is proportional to the area of the pulse, within the limits of a frequency inverval where the period remains large in comparison with the duration of the pulse.[1]

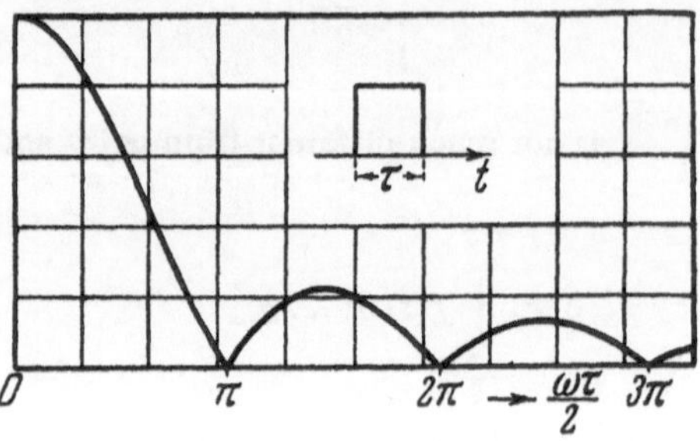

Fig. 19.

We shall have more than one further occasion to turn to this criterion.

With an increase in frequency, in which case T is made comparable with τ, the function S begins to decrease. Before examining the general behavior, we shall consider a number of examples.

Let us compute the spectrum of a rectangular pulse of height $\underline{h}$ and duration τ. We obtain

$$S = h \int_{-\frac{\tau}{2}}^{+\frac{\tau}{2}} e^{-j\omega t}\, dt = -\frac{h}{j\omega} e^{-j\omega t} \Big|_{-\frac{\tau}{2}}^{+\frac{\tau}{2}} = q \frac{\sin \omega \frac{\tau}{2}}{\omega \frac{\tau}{2}}, \qquad (11.4)$$

where the area under the pulse is $q = h\tau$. A graph of $\Phi = |S|$ for this case is is illustrated in Fig. 19.

As an illustration of the preceding, we note that when $\omega \frac{\tau}{2} \ll 1$

$$\sin \omega \frac{\tau}{2} \approx \omega \frac{\tau}{2}$$

[1]One fine point should be mentioned here: the formulated criterion is valid for a pulse that does not change sign. The details relevant to this are developed in Appendix IV at the end of the book.

and

$$S \approx q.$$

For a pulse in the form of a triangle with base τ and height h we have

$$f(t) = \begin{cases} 0 & t < -\frac{\tau}{2}, \\ h\left(1 + \frac{2t}{\tau}\right) & -\frac{\tau}{2} < t < 0, \\ h\left(1 - \frac{2t}{\tau}\right) & 0 < t < \frac{\tau}{2}, \\ 0 & \frac{\tau}{2} < t. \end{cases}$$

Consequently,

$$S = h \int_{-\frac{\tau}{2}}^{0} \left(1 + \frac{2t}{\tau}\right) e^{-j\omega t}\, dt + \\ + h \int_{0}^{\frac{\tau}{2}} \left(1 - \frac{2t}{\tau}\right) e^{-j\omega t}\, dt = q \frac{1 - \cos \omega \frac{\tau}{2}}{\frac{1}{2}\left(\omega \frac{\tau}{2}\right)^2}, \tag{11.5}$$

where $q = \frac{1}{2} h\tau$ (Fig. 20).

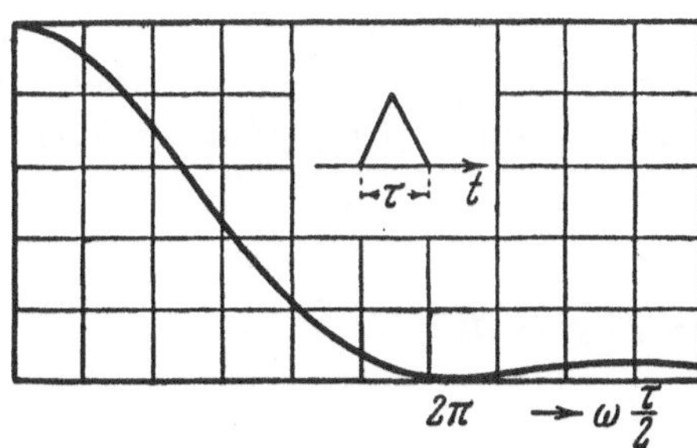

Fig. 20.

Expanding the numerator into a power series, we can be sure that the ratio will tend to unity as $\omega \frac{\tau}{2}$ approaches zero. For a cosine pulse, which is a section of a cosine wave with period 2τ and amplitude h, we have

$$S = h \int_{-\frac{\tau}{2}}^{+\frac{\tau}{2}} e^{-j\omega t} \cos \pi \frac{t}{\tau} dt = q \frac{\cos \omega \frac{\tau}{2}}{1 - \left(\frac{2}{\pi} \omega \frac{\tau}{2}\right)^2}, \qquad (11.6)$$

where $q = \frac{2}{\pi} h\tau$ (Fig. 21).

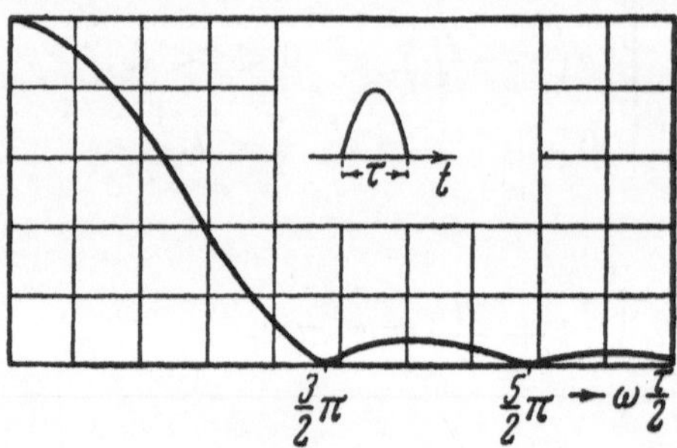

Fig. 21.

In more recent times attention has been drawn to a pulse, which in the specialized literature is called a "bell" pulse. This pulse is expressed by the function

$$f(t) = e^{-\beta^2 t^2}.$$

This function is remarkable in that it duplicates itself, i.e., its Fourier transform gives the same function. In fact, for the spectrum of a bell pulse we obtain

$$S = \int_{-\infty}^{+\infty} e^{-\beta^2 t^2} e^{-j\omega t} dt = e^{-\frac{\omega^2}{4\beta^2}} \int_{-\infty}^{+\infty} e^{-\left(\beta t + j\frac{\omega}{2\beta}\right)^2} dt =$$

$$= \frac{2}{\beta} e^{-\frac{\omega^2}{4\beta^2}} \int_{0}^{\infty} e^{-x^2} dx = \frac{\sqrt{\pi}}{\beta} e^{-\frac{\omega^2}{4\beta^2}}$$

(Fig. 22).

Let us now consider a group of pulses, where the origin is at $t = 0$, but where the amplitude tends to zero when and only when $t \to \infty$.

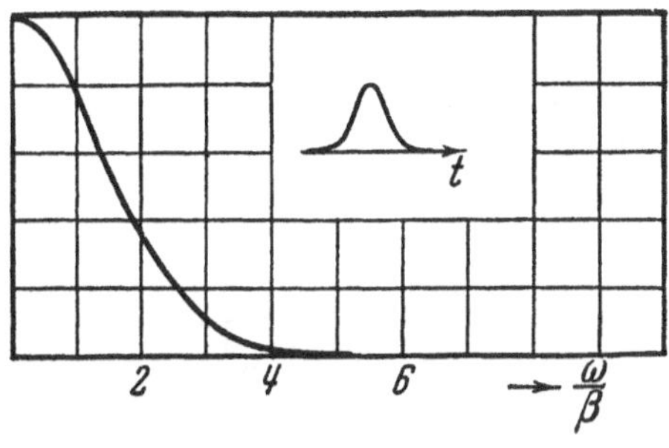

Fig. 22.

Let, for example, an exponential pulse be specified:

$$f(t) = e^{-\beta t}\sigma_0(t).$$

Its spectrum is

$$S = \int_0^\infty e^{-(\beta+j\omega)t}\,dt = \frac{e^{-(\beta+j\omega)t}}{\beta+j\omega}\Bigg|_0^\infty = \frac{1}{\beta+j\omega} \tag{11.7}$$

(Fig. 23).

The spectrum of a pulse in the form of a decaying sinusoid

$$f(t) = e^{-\alpha t}\sin\omega_1 t\sigma_0(t)$$

has the form

$$S = \int_0^\infty e^{-(\alpha+j\omega)t}\sin\omega_1 t\,dt = \frac{\omega_1}{\alpha^2-\omega^2+\omega_1^2+2j\alpha\omega}$$

or, introducing the notation

$$\alpha^2+\omega_1^2 = \omega_0^2, \quad d = \frac{2\alpha}{\omega_0},$$

$$S = \frac{\omega_1}{\omega_0^2}\,\frac{1}{1-\left(\frac{\omega}{\omega_0}\right)^2+jd\frac{\omega}{\omega_0}} \tag{11.8}$$

(Fig. 24). Let us further compute the spectrum of a pulse in the form of a section from a sinusoid, where the section consists of a whole number of periods n:

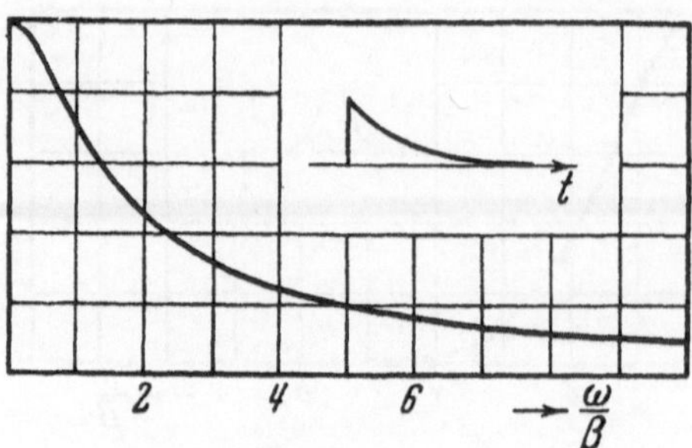

Fig. 23.

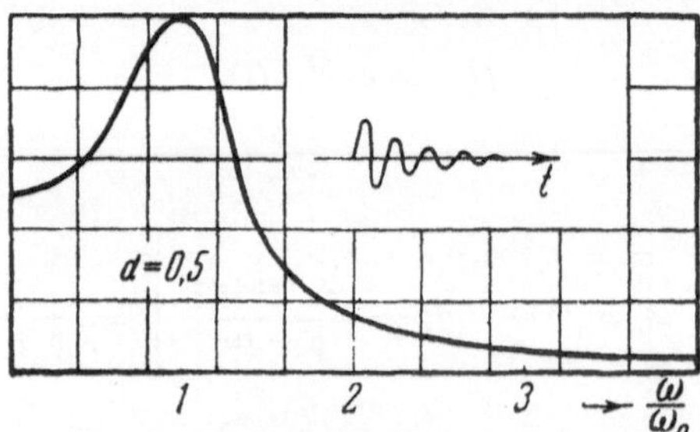

Fig. 24.

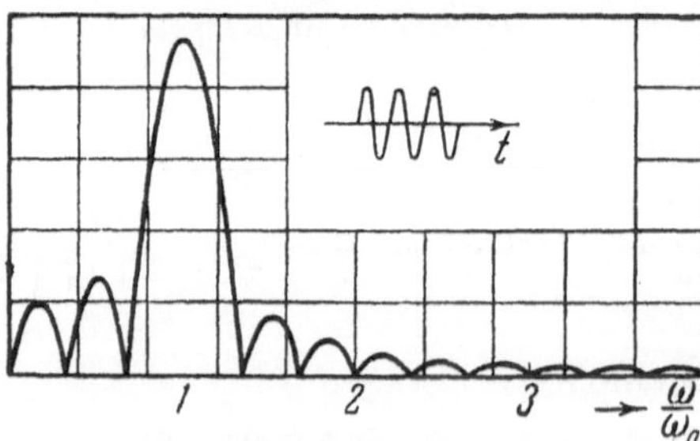

Fig. 25.

$$S = \int_{-\frac{nT}{2}}^{+\frac{nT}{2}} e^{-j\omega t} \sin \omega_0 t \, dt = \frac{2j\omega_0}{\omega_0^2 - \omega^2} (-1)^n \sin n\pi \frac{\omega}{\omega_0} \qquad (11.9)$$

(Fig. 25). This is essentially nothing more than the running spectrum of a sinusoid (see Section 5).

Very often pulses of a definite form are periodically repeated. Let us establish the connection between the spectra of a single pulse and a periodic succession of the same pulses. One thing is clear beforehand: the spectrum of the single pulse is a complex spectrum, since the pulse is a nonperiodic function. If now a pulse of any arbitrary form is periodically repeated, then we obtain a periodic function, which has a discrete harmonic spectrum.

Let the spectrum of the single pulse be

$$S_0 = \int_{-\infty}^{+\infty} e^{-j\omega t} f(t)\, dt \,. \tag{11.10}$$

If such a pulse is repeated every time interval T, then a periodic function with period T is obtained (Fig. 26). The spectrum of this function can be obtained from the equation

$$C_k = \frac{1}{T} \int_{-\frac{T}{2}}^{+\frac{T}{2}} e^{-j2\pi k \frac{t}{T}} f(t)\, dt \,. \tag{11.11}$$

Comparing (11.10) and (11.11), we see that the values of the continuous function S_0 coincide with the values of C_k (except for the constant multiplier $1/T$) at definite values of the argument, namely, when

$$\omega = 2\pi \frac{k}{T} = k\omega_1,$$

where $\omega_1 = \dfrac{2\pi}{T}$ is the circular frequency of repetition.

Therefore, the set of points TC_k, which define the discrete spectrum of the periodic succession of pulses, lies on the curve S_0 defining the spectrum of the single pulse.

It may be additionally stated that the line spectrum of a periodic succession of pulses is contained in the curve of the complex spectrum of one of the pulses alone.

In this example it is easy to trace the limiting transition from the Fourier

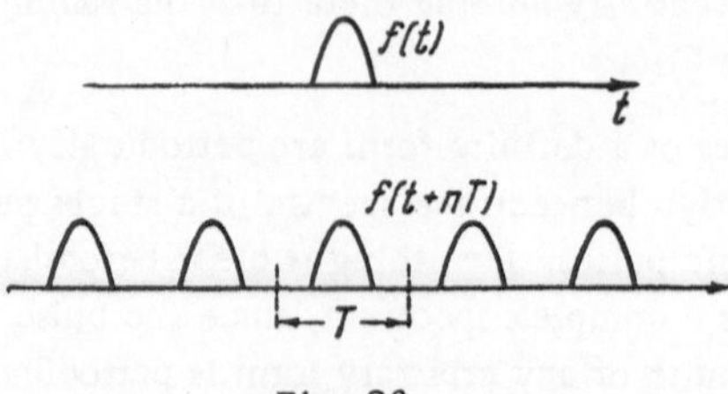

Fig. 26.

series to the Fourier integral: if the period of repetition increases, i.e., if the pulses are repeated less and less frequently, then the points representing the line spectrum remain on the curve for S_0 and come closer and closer together, until a continuous succession, i.e., a curve, coinciding with S_0 is formed.

12. Relation Between Pulse Duration and Width of Spectrum

We have already established the fact that the shorter the pulse, the broader its spectrum will be, in particular, a pulse of infinitesimal duration has an infinitely broad spectrum with a uniform density. In this connection there appears one relationship, which is quite general and has very great significance, and which we shall proceed to establish step by step.

We note first of all that a general idea of the connection between the extensiveness in time and with respect to frequency follows directly from a general property of the Fourier transform. Let us assume that the function f(t) has the spectrum S(ω). Let us change the time scale by a factor a, then determine the spectrum of the function af(at) (the factor a in front of the function is added in order to preserve the area, i.e., the dimensions of the graph of the function are increased along the ordinate in the same proportion as along the abscissa):

$$S_a = a\int_{-\infty}^{+\infty} f(at)e^{-j\omega t}\,dt = S\left(\frac{\omega}{a}\right).$$

Thus, if the duration of the function f(t) is reduced by a factor of a, then the spectrum is increased by exactly the same proportion. In this case it is assumed that the definitions of the duration of the function and the width of its spectrum remain fixed.

Now let us consider a few examples.

Let us once again look at the rectangular pulse [see Section 11, Eq. (11.4) and Fig. 18]. For the spectrum of such a pulse we obtained

$$S = q\frac{\sin \omega \frac{\tau}{2}}{\omega \frac{\tau}{2}}. \tag{12.1}$$

Let us now compare the pulse duration and width of the spectrum. In this case the pulse duration Δt is naturally understood to be the quantity τ. As for the width of the spectrum, here it is necessary to choose some definition, since the spectrum of the pulse is unbounded. But since the spectrum decays with frequency it is possible, for example, to agree upon the frequency band Δf between zero and that value at which the spectrum first returns to zero as the spectrum width. From Eq. (12.1) it is evident that this does take place when the argument of the sine becomes equal to π. Hence we get the following:

$$2\pi\Delta f\,\frac{\Delta t}{2} = \pi$$

or

$$\Delta f\Delta t = 1,$$

i.e., the product of the duration of the given pulse Δt and the width of its spectrum Δf (according to the arbitrary definition that we have chosen) is equal to unity.

For a triangular pulse we obtained

$$S = q\frac{1 - \cos \omega\frac{\tau}{2}}{\frac{1}{2}\left(\omega\frac{\tau}{2}\right)^2}. \tag{12.2}$$

The first zero of the spectrum will occur at

$$\omega\frac{\Delta t}{2} = 2\pi,$$

whence

$$\Delta f\Delta t = 2.$$

We shall retain the definitions for Δf and Δt: Δf is the frequency band to the first point where the spectrum passes through zero, Δt is the interval outside of which the function representing the pulse is equal to zero.

For a cosine pulse

$$S = q\frac{\cos \omega \frac{\tau}{2}}{1-\left(\frac{2}{\pi}\omega\frac{\tau}{2}\right)^2}, \tag{12.3}$$

and the relation we are seeking assumes the form

$$\Delta f \Delta t = 1{,}5.$$

For all of the examples considered it is found that

$$\Delta f \Delta t \approx 1. \tag{12.4}$$

However, if we should desire to increase the number of examples, we would very quickly encounter difficulty in choosing the definition of Δf and Δt for every concrete example. Let us assume, for example, that we wish to establish a relation of the type (12.4) for the exponential pulse

$$f(t) = e^{-\beta t}\sigma(t).$$

The previous definition of Δt is unsuitable, since it is obvious that this function is not equal to zero anywhere in the interval from 0 to ∞. The question naturally arises as to how one might define the duration in some universal way.

We note first of all that the problem of the relationship between Δf and Δt is not only of theoretical, but also great practical significance. In modern pulse techniques it is required that very short, but at the same time very power ful pulses be developed. In other words, a very high-energy pulse must be concentrated into a very short interval of time.

On the other hand, it is required that the spectrum of the pulse be, insofar as possible, flattened out very little, since a broad spectrum elicits a whole series of serious difficulties in tuning the pulsing apparatus.

Thus, on the one hand, we require a small value for Δt, while, on the other hand, we must have a small value of Δf. These requirements, as we have seen, generally speaking, are mutually contradicting. However, it is possible to find a particular pulse shape for which the product $\Delta f \Delta t$ has a least value.

If we approach the problem of defining the quantities Δf and Δt from the practical point of view, then we can assume the following definition of the duration: the duration of a pulse is understood to mean that interval of time in which the greater part of the pulse energy is concentrated. Analytically, this definition can be formulated as follows:

$$\int_{t_0-\frac{\Delta t}{2}}^{t_0+\frac{\Delta t}{2}} f^2(t)\,dt = \eta \int_{-\infty}^{+\infty} f^2(t)\,dt = \varepsilon A_t \,. \tag{12.5}$$

Here

$$A_t = \int_{-\infty}^{+\infty} f^2(t)\,dt$$

is a quantity which is proportional to the total energy of the pulse; η is a proper fraction expressing the relative part of the total energy of the pulse belonging to the time interval Δt. Equation (12.5) can be conveniently solved by means of planimetric integration.

The width of the spectrum can be defined analogously:

$$\int_0^{\Delta\omega} \Phi^2(\omega)\,d\omega = \eta \int_0^{\infty} \Phi^2(\omega)\,d\omega = \varepsilon A_\omega \,. \tag{12.6}$$

We note that according to the Rayleigh theorem

$$A_\omega = \pi A_t \,. \tag{12.7}$$

As for the quantity t_0 appearing in the limits of the integral on the left-hand side (12.5), in a number of instances the choice of this quantity leaves no doubt. For symmetic pulses, expressed by even functions, $t_0 = 0$. For pulses that start at $t = 0$, Eq. (12.5) would be rewritten in the form

$$\int_0^{\Delta t} f^2(t)\,dt = \eta \int_0^{\infty} f^2(t)\,dt \,.$$

Let us turn now to the pulses already considered and compute their durations and widths of their spectra, basing our procedure on the adopted general definition. Choosing for η the value 0.9, we obtain the following tabulated summary of the quantities of interest to us (the details of the calculations are given in Appendix VI at the end of the book).

We shall concern ourselves only with one remark relative to this table: as we can see, $\Delta f \Delta t$ turns out to be greatest for pulses characterized by a discontinuity in the function f(t) (rectangular and exponential); the smallest value of $\Delta f \Delta t$ is obtained for pulses with a discontinuity in the first derivative (triangular and cosine) and, finally, the least value of $\Delta f \Delta t$ turns up for the bell pulse, which is distinguished by the fact that the function expressing it is discontinuous in all of its derivatives.

Pulse	Fig.	Δt	$\Delta\omega$	Δf	$\Delta f\,\Delta t$
Rectangular	19	$0.90\,\tau$	$5.1\,\frac{1}{\tau}$	$0.81\,\frac{1}{\tau}$	0.73
Triangular	20	$0.541\,\tau$	$5.3\,\frac{1}{\tau}$	$0.84\,\frac{1}{\tau}$	0.46
Cosine	21	$0.596\,\tau$	$4.57\,\frac{1}{\tau}$	$0.73\,\frac{1}{\tau}$	0.43
Bell	22	$0.825\,\frac{1}{\beta}$	$1.64\,\beta$	$0.26\,\beta$	0.22
Exponential	23	$1.155\,\frac{1}{\alpha}$	$6.16\,\alpha$	$0.98\,\alpha$	1.13

From all that has been outlined we can conclude that in the general case the connection between Δf and Δt satisfies the inequality

$$\Delta f\,\Delta t \geqslant \mu, \tag{12.8}$$

where μ is some constant depending, of course, on the choice of definition of Δf and Δt.

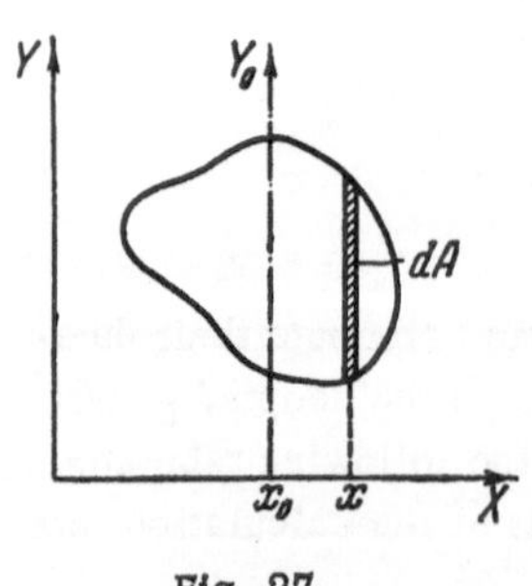

Fig. 27.

The definitions that we have just applied, for all their simplicity and practical convenience, unfortunately do not yet make it possible to resolve the problem in general form. We shall introduce new definitions for Δf and Δt, based on the application of the concept of functional moments.

For greater visualization we shall explain these definitions with reference to the generally known concepts and definitions of applied mechanics. Let us assume that we have been given an arbitrary plane figure in the XY-plane, and that we are required to determine the dimension of this figure in the direction of the X-axis in some universal way (Fig. 27). Such a universal measure could be fulfilled by the radius of inertia of the given figure relative to the principal axis Y_0, which is parallel to Y.

Let us recall the definitions. The moment of inertia of a plane figure relative to the Y-axis is the integral

$$I = \int x^2 \, dA.$$

The principal axis is the name given to the axis passing through the areal center of mass. For determining the coordinate of the center of mass, i.e., the position of the principal axis, it is necessary to find the so-called static moment of the area

$$M = \int x \, dA,$$

after which the coordinate of the center of mass is determined from the equation

$$x_0 = \frac{\int x \, dA}{\int dA} = \frac{M}{A}.$$

The principal moment of inertia, i.e., the moment relative to the principal axis Y_0, will be

$$I_0 = \int (x - x_0)^2 \, dA = \int x^2 \, dA - 2x_0 \int x \, dA + x_0^2 \int dA =$$
$$= I + x_0^2 A - 2x_0 M = I - x_0^2 A = I - \frac{M^2}{A}.$$

If we assume

$$I_0 = r^2 A,$$

then in this definition the quantity r will then be the so-called radius of inertia.

Thus

$$r^2 = \frac{I}{A} - \frac{M^2}{A^2}, \tag{12.9}$$

and the quantity defined in this way can serve as a universal measure of the width of the given figure in the direction of the X-axis.

Let us turn now to pulses and their spectra. The graphs of all pulses and their spectra, including those already mentioned, are plane figures to which the above is fully applicable. For pulses the time axis plays the role of the X-axis, for spectra the same part is taken by the frequency axis. An element of area of the figure is expressed by the product of the value of the function by the differential of the independent variable. However, in view of the fact that we shall come across functions which change sign, it will be more convenient to operate, not on the function itself, but on its square.

Thus we can introduce the following notation and definitions:

$$A_t = \int_{-\infty}^{+\infty} f^2(t)\,dt, \qquad A_\omega = \int_0^\infty \Phi^2(\omega)\,d\omega,$$

$$M_t = \int_{-\infty}^{+\infty} tf^2(t)\,dt, \qquad M_\omega = \int_0^\infty \omega\Phi^2(\omega)\,d\omega,$$

$$I_t = \int_{-\infty}^{+\infty} t^2f^2(t)\,dt, \qquad I_\omega = \int_0^\infty \omega^2\Phi^2(\omega)\,d\omega.$$

Furthermore, according to (12.9),

$$\Delta t^2 = \frac{I_t}{A_t} - \frac{M_t^2}{A_t^2}; \quad \Delta\omega^2 = \frac{I_\omega}{A_\omega} - \frac{M_\omega^2}{A_\omega^2}.$$

From this we get

$$(\Delta\omega\,\Delta t)^2 = \left(\frac{I_\omega}{A_\omega} - \frac{M_\omega^2}{A_\omega^2}\right)\left(\frac{I_t}{A_t} - \frac{M_t^2}{A_t^2}\right)$$

or, making use of (12.7) and introducing $\Delta f = \frac{1}{2\pi}\Delta\omega$,

$$(\Delta f\,\Delta t)^2 = \frac{1}{4\pi^4 A_t^4}(I_\omega A_\omega - M_\omega^2)(I_t A_t - M_t^2). \tag{12.10}$$

Now let us qualify and simplify the problem, assuming that we are concerned with even functions of time. Then

$$\Phi(\omega) = a(\omega) = \int_{-\infty}^{+\infty} f(t) \cos \omega t \, dt$$

and, moreover,

$$t_0 = 0; \quad M_t = 0.$$

Let us agree, in addition, that the function f(t) is normalized, i.e., that

$$A_t = \int_{-\infty}^{+\infty} f^2(t) \, dt = 1$$

and, correspondingly,

$$A_\omega = \int_{-\infty}^{+\infty} \Phi^2(\omega) \, d\omega = \pi.$$

Let us now take the integral

$$I_t = \int_{-\infty}^{+\infty} t^2 f^2(t) \, dt.$$

We have

$$a(\omega) = \int_{-\infty}^{+\infty} f(t) \cos \omega t \, dt.$$

Differentiating both sides with respect to ω,

$$a'(\omega) = -\int_{-\infty}^{+\infty} t f(t) \sin \omega t \, dt = -b_1(\omega),$$

where $b_1(\omega)$ is the spectrum of the odd function

$$f_1(t) = t f(t).$$

Applying the Rayleigh theorem, we obtain

$$I_t = \int_{-\infty}^{+\infty} t^2 f^2(t)\, dt = \frac{1}{\pi} \int_0^{\infty} a'^2(\omega)\, d\omega = \frac{1}{\pi} N.$$

Introducing all these relations into (12.10), we find (from here on we shall omit writing the variable ω)

$$(\Delta f \Delta t)^2 = \frac{1}{4\pi^5} N(\pi I - M^2). \tag{12.11}$$

Let us now attempt to find the least value of $\Delta f \Delta t$, i.e., to pick out that function which will give the minimum value of the integral expression

$$K = N(\pi I - M^2). \tag{12.12}$$

For this purpose we shall apply variational methods, following the example of Maier and Leontovich [11].

Let us write the first variation of the quantity (12.12):

$$\delta K = \delta N(\pi I - M^2) + N(\pi \delta I - 2M\, \delta M).$$

In order to find the minimum we should equate this variation to zero. But here the integrals I, M, and N stand for fully defined constant values, which we shall designate by $\underline{i}$, $\underline{m}$, and $\underline{n}$, respectively. After this we can combine the varied integrals and write

$$\delta \int_0^{\infty} [(\pi i - m)\ (a')^2 + (\pi n \omega^2 - 2mn\omega)\, a^2]\, d\omega =$$

$$= \delta \int_0^{\infty} F(\omega,\ a, a')\, d\omega = 0.$$

We can now write the Euler equation

$$\frac{\partial F}{\partial a} - \frac{\partial}{\partial \omega}\left(\frac{\partial F}{\partial a'}\right) = 0$$

in the following form:

$$(\pi i - m^2)\, a'' - (\pi n \omega^2 - 2mn\omega)\, a = 0. \tag{12.13}$$

Let us multiply this equation by $\underline{a}$ and integrate from zero to infinity. The integral in the first term is taken by parts:

$$\int_0^\infty aa''\,d\omega = aa'\Big|_0^\infty - \int_0^\infty (a')^2\,d\omega = -n$$

[since $a(\infty) = a'(0) = 0$]. Integrating the second term, we note that

$$\int_0^\infty \omega^2 a^2\,d\omega = i; \qquad \int_0^\infty \omega a^2\,d\omega = m,$$

so that we consequently obtain the relation

$$2\pi i = 3m^2.$$

Noting that, according to (12.12),

$$K_{\min} = n(\pi i - m^2),$$

we can now write

$$K_{\min} = \frac{\pi}{3} ni = \frac{\pi}{3} \int_0^\infty (a')^2\,d\omega \cdot \int_0^\infty \omega^2 a^2\,d\omega.$$

In order to evaluate this expression we shall refer to the Bunyakovskii inequality

$$\int_a^b f_1^2(x)\,dx \cdot \int_a^b f_2^2(x)\,dx \geqslant \left[\int_a^b f_1(x) f_2(x)\,dx\right]^2.$$

In the present case

$$\int_0^\infty (a')^2\,d\omega \cdot \int_0^\infty \omega^2 a^2\,d\omega \geqslant \left[\int_0^\infty \omega a a'\,d\omega\right]^2.$$

Computing the integral on the right-hand side of the inequality by parts, we find that it is equal to $-\frac{\pi}{2}$. Therefore,

$$K_{\min} \geqslant \frac{1}{12}\pi^3$$

and

$$(\Delta f\,\Delta t)^2 \geqslant \frac{1}{48\pi^2},$$

from which, finally,

$$\Delta f\,\Delta t \geqslant \frac{1}{4\sqrt{3\pi}} = 0.046.$$

We have thus obtained an evaluation of $\Delta f \Delta t$ for the most advantageous case. It is interesting to compare the quantity so obtained with the value of $\Delta f \Delta t$ for a bell pulse, which, as previously seen, has the lowest value for the product $\Delta f \Delta t$ of all the pulses that we have considered. It is understandable that the data computed earlier cannot possibly be used, since we would immediately be applying other criteria altogether. We have to compute the integral appearing in Eq. (12.11).

The bell pulse and its spectrum have been determined above (Section 11):

$$f(t) = e^{-\beta^2 t^2}; \qquad a(\omega) = \frac{\sqrt{\pi}}{\beta}\, e^{-\frac{\omega^2}{4\beta^2}}.$$

From the condition for normalization,

$$A_t = \int_{-\infty}^{+\infty} f^2(t)\,dt = 1$$

we determine $\beta = \sqrt{\frac{\pi}{2}}$. After this we find

$$I = \int_0^\infty \omega^2 a^2\,d\omega = 2\int_0^\infty e^{-\frac{\omega^2}{\pi}}\,\omega^2\,d\omega = \frac{\pi^2}{2},$$

$$N = \int_0^\infty a'^2\,d\omega = \frac{2}{\pi^2}\int_0^\infty e^{-\frac{\omega^2}{\pi}}\,\omega^2\,d\omega = \frac{1}{2},$$

$$M = \int_0^\infty \omega a^2\,d\omega = 2\int_0^\infty e^{-\frac{\omega^2}{\pi}}\,\omega\,d\omega = -\pi.$$

Comparing these values in Eq. (12.12), we obtain

$$\Delta f \Delta t = \frac{1}{4\pi}\sqrt{\frac{\pi-2}{\pi}} = 0.048 .$$

This quantity, as we can see, is very near the theoretical minimum.

13. Relation Between the Spectra and Characteristics of a Linear System

Let a linear system be specified and let it be described by an ordinary nth-order differential equation with constant coefficients:

$$a_n\frac{d^n y}{dt^n} + a_{n-1}\frac{d^{n-1} y}{dt^{n-1}} + \ldots + a_1\frac{dy}{dt} + a_0 y = x(t). \tag{13.1}$$

Let us apply the Fourier transform to both sides of the equation:

$$\int_{-\infty}^{+\infty} f(t)\, e^{-j\omega t}\, dt.$$

On the right-hand side the spectrum of the function x(t) is obtained; when integrating the left-hand side we apply Eq. (4.3). Then

$$[a_n(j\omega)^n + a_{n-1}(j\omega)^{n-1} + \ldots + a_1 j\omega + a_0]\, S_y = S_x \tag{13.2}$$

or, introducing abbreviated notation for the polynomial in the brackets,

$$Z(\omega)\, S_y = S_x;$$

here S_x and S_y are the spectra of the functions x(t) and y(t), respectively. Thus

$$S_y = \frac{S_x}{Z}, \tag{13.3}$$

and we can compute y from the equation

$$y(t) = \frac{1}{2\pi}\int_{-\infty}^{+\infty} \frac{S_x}{Z}\, e^{j\omega t}\, d\omega. \tag{13.4}$$

This equation gives a solution to Eq. (13.1) by the method of Fourier integrals.

Let us further introduce the notation

$$K(\omega) = \frac{1}{Z(\omega)}.$$

This quantity, which expresses the ratio of the complex amplitudes at the output and input of the system when operating with a sinusoid, is called the complex frequency characteristic.

We can now rewrite (13.3) in the form

$$S_y = KS_x. \tag{13.5}$$

Let us now consider pulse excitation of the system. Let the following unit pulse be delivered to the input:

$$x(t) = \delta(t).$$

In this case at the output a certain functional reaction will be observed as a result of the action of the system; this we can designate

$$y(t) = g(t).$$

The function g(t) is called the time characteristic of the system.[1]

For the unit pulse $S_x = 1$; consequently, in this case, according to (12.5),

$$S_y = K,$$

so that

$$g(t) = \frac{1}{2\pi} \int_{-\infty}^{+\infty} K(\omega) e^{j\omega t} d\omega$$

and

$$K(\omega) = \int_{-\infty}^{+\infty} g(t) e^{-j\omega t} dt, \tag{13.6}$$

[1]Alternate terminologies are possible: g(t) is also called the pulse reaction; $h(t) = \int_0^t g(\tau)\, d\tau$ is called the transfer function or transfer characteristic.

i.e., the complex frequency characteristic of a linear system is the spectrum of the system time characteristic.

Let us take an example. Let an electrical circuit consisting of the elements R, L, and C, connected in series, be specified. The differential equation of such a system has the form

$$L\frac{di}{dt}+Ri+\frac{1}{C}\int i\,dt=e(t)\,.$$

Let us apply the notation

$$U=\frac{1}{C}\int i\,dt;$$

then the equation assumes the form

$$\frac{d^2U}{dt^2}+2\alpha\frac{dU}{dt}+\omega_0^2U=\omega_0^2e(t),$$

where, as usual,

$$2\alpha=\frac{R}{L};\quad \omega_0^2=\frac{1}{LC}.$$

In the present case

$$K(\omega)=\frac{\omega_0^2}{\omega_0^2-\omega^2+2\alpha j\omega}\,. \tag{13.7}$$

The time characteristic is obtained as the solution of the equation

$$\frac{d^2g}{dt^2}+2\alpha\frac{dg}{dt}+\omega_0^2g=\omega_0^2\delta(t)$$

or, in operational form,

$$\bar{g}=pK(p),$$

whence

$$g(t)=\frac{\omega_0^2}{\omega_1}e^{-\alpha t}\sin\omega_1 t, \tag{13.8}$$

where

$$\omega_1^2 = \omega_0^2 - \alpha^2.$$

We see the Eq. (13.7) does in fact give the spectrum of the function (13.8) [compare with Eq. (11.8)].

From (13.6) there follows the possibility of experimentally obtaining the frequency characteristic of a given system by means of analysis. In fact, instead of directly recording the frequency characteristic in connection with sinusoidal excitation of the system investigated, it is possible to excite the system with a short pulse and to analyze the resultant time function at the system output. Rimskiĭ-Korsakov and Shumova [13] made use of this possibility for acoustical measurements; they excited the object of investigation with periodically repeated short pulses and obtained the desired frequency characteristic in the form of a line spectrum by means of an automatic analyzer.

We began our discussion with the case of a system with lumped constants, which was described by an equation in total derivatives. However, the relation (13.3) [or (13.5)] is also valid in the case when the system is described by partial differential equations. In the latter case Z and K are usually transcendental functions of ω.

In concluding this section, we note, in addition, that as a result of (13.6) the following relation is applicable to the characteristics of a linear system:

$$\Delta f \, \Delta t \geqslant \mu.$$

In the present application it is considered as follows: the product of the duration Δt of the time characteristic and the width Δf of the frequency characteristic for any linear system is greater than some constant. In this case it is understood that appropriate definitions are chosen for Δf and Δt (see Section 12).

14. Functions with Bounded Spectra

When dealing with the transmission of various signals in communications engineering we are usually concerned with time functions whose spectra are bounded, i.e., where the spectrum contains no frequencies above a certain limit. Such functions have one remarkable property, which was first established in 1933 by V. A. Kotel'nikov [8] and which was expressed by him in a theorem that played a fundamental role in the theory of communications and, in particular, in pulse techniques.

This property is contained in the fact that, while in the general case the time function is fully defined by an innumerable set of values (i.e., an infinite number of values over a finite interval), a function with a bounded spectrum is fully defined by a denumerable set of values (i.e., by a finite number of values over a finite interval). From the geometric point of view this means that if a fully defined set of points representing the multiple values of a function with a bounded spectrum is specified over a finite interval, then a continuous curve representing the graph of the function can be drawn through these points in one-to-one correspondence. This assumption is explained by the fact that the absence of the higher frequencies in the composition of the function imposes a rigid limitation on the ways in which any two adjacent points can be joined, and the meaning of Kotel'nikov's theorem is precisely contained in the assertion that for a sufficiently frequent distribution of points these limitations lead to the curve being completely defined by these points.

This idea can be further clarified in the following example. Let us assume that, relevant to a continuous line, it is known that it is a broken line (in the present example this, then, is a limitation on the properties of the line), and let us assume that we have specified the points of discontinuity for this line. Then, it being known from the property of a broken line that the points of discontinuity are joined by straight line segments, we can easily reproduce the entire broken line.

Let us now proceed to a proof of the theorem, which in the author's formulation was expressed: "Any function f(t) consisting of frequencies from 0 to f_c can, with any desired accuracy be treated as a succession of numbers recurring every $\frac{1}{2f_c}$ seconds."

The proof consists in expanding the function in a special series form.

In the general case

$$f(t) = \frac{1}{2\pi} \int_{-\infty}^{+\infty} S(\omega)\, e^{j\omega t}\, d\omega, \tag{14.1}$$

where

$$S(\omega) = \int_{-\infty}^{+\infty} f(t)\, e^{-j\omega t}\, dt. \tag{14.2}$$

However, in the special case considered, of a function with a bounded spectrum, we have instead of (14.1)

$$f(t)=\frac{1}{2\pi}\int_{-\omega_c}^{\omega_c} S(\omega)\, e^{j\omega t}\, d\omega, \tag{14.3}$$

since $S(\omega) = 0$ for $\omega > \omega_c$. However, the function $S(\omega)$ over the finite interval $(-\omega_c, \omega_c)$ can be expanded in a Fourier series with respect to the frequencies as follows:

$$S(\omega)=\sum_{-\infty}^{+\infty} D_k e^{j\pi k\frac{\omega}{\omega_c}}, \tag{14.4}$$

where $2\omega_c$ acts as the period with respect to frequency. The coefficients D_k of the series (14.4) are determined from the equation

$$D_k=\frac{1}{2\omega_c}\int_{-\omega_c}^{\omega_c} S(\omega)\, e^{-j\pi k\frac{\omega}{\omega_c}}\, d\omega. \tag{14.5}$$

Substituting (14.4) into (14.3), we obtain

$$f(t)=\frac{1}{2\pi}\int_{-\omega_c}^{\omega_c}\left(\sum_{-\infty}^{+\infty} D_k e^{j\pi k\frac{\omega}{\omega_c}}\right) e^{j\omega t}\, d\omega$$

or, inverting the order of the operations,

$$f(t)=\frac{1}{2\pi}\sum_{-\infty}^{+\infty} D_k \int_{-\omega_c}^{\omega_c} e^{j\omega\left(t+k\frac{\pi}{\omega_c}\right)}\, d\omega,$$

from which, after integration, we find

$$f(t)=\frac{1}{\pi}\sum_{-\infty}^{+\infty} D_k \frac{\sin \omega_c\left(t+k\frac{\pi}{\omega_c}\right)}{t+k\frac{\pi}{\omega_c}}. \tag{14.6}$$

Now let us compare (14.5) and (14.3). As we can see,

$$D_k=\frac{\pi}{\omega_c} f\left(-k\frac{\pi}{\omega_c}\right)=\Delta t\cdot f(-k\,\Delta t).$$

Substituting the value that has been found for D_k into (14.6) and changing the sign after the summation sign (since the summation is performed over all values of k from $-\infty$ to $+\infty$), we obtain finally

$$f(t)=\sum_{-\infty}^{+\infty} f(k\,\Delta t)\,\frac{\sin\omega_c(t-k\,\Delta t)}{\omega_c(t-k\,\Delta t)}. \tag{14.7}$$

Thus a function with a bounded spectrum can be represented by the series (14.7), the coefficients of which serve as readings of the values of the function taken every interval

$$\Delta t=\frac{\pi}{\omega_c}=\frac{1}{2f_c}.$$

The theorem is thus proved.

In the preceding proof readings of multiple values of the function with a bounded spectrum acted as numbers defining the function. But these could just as well have been other independent numbers, as shown by the discussion which follows.

Every function (with some unessential reservations) can be represented over an arbitrary finite interval $\left(-\frac{T}{2},\frac{T}{2}\right)$ by the trigonometric series

$$f(t)=\sum_{-\infty}^{\infty} C_k e^{j2\pi k\frac{t}{T}}. \tag{14.8}$$

But if the spectrum of the function f(t) is bounded, i.e., if there are no frequencies appearing in its composition above the limiting frequency ω_c, then the expansion (14.8) is also restricted to a finite number of components, and we have in such a case

$$f(t)=\sum_{-n}^{n} C_k e^{j2\pi k\frac{t}{T}}, \tag{14.9}$$

where the number n of the highest harmonic is determined from the relation

$$2\pi\frac{n}{\tau}=\omega_c$$

or

$$n=f_c\tau.$$

Therefore, the function is completely defined over the interval τ through n complex amplitudes C_k and the constant component C_0. But, since each amplitude C_k is in turn defined by two numbers (or, what amounts to the same thing, each component of the polynomial (14.9) is defined by an amplitude and a phase), the total number necessary for completely defining the function over the interval τ is

$$m = 2n + 1 = 2f_c\tau + 1. \tag{14.10}$$

Now it is necessary to let τ tend to infinity, in order to determine the function f(t) over the whole time axis. Here, of course, n will tend to infinity, so that in place of (14.10) we shall have the asymptotic expression

$$m \sim 2n = 2f_c\tau.$$

If we are to transfer the numbers uniformly, then per unit time it is necessary to transfer

$$\frac{m}{\tau} = 2f_c$$

numbers, while the time interval between two frequencies will be

$$\Delta t = \frac{\tau}{m} = \frac{1}{2f_c}.$$

Thus it has been shown that the numbers defining a function with a bounded spectrum can serve not only as readings of multiple values of the function, but also coefficients of its expansion in a Fourier series.

In practical application we are concerned with processes having a beginning and an end, i.e., with functions having a limited duration T. Time-bounded functions cannot have a bounded spectrum (i.e., a spectral density equal to zero outside some finite interval), as these conditions are mutually contradictory. However, it is possible to formulate the discussion on a more general basis, defining the spectrum width F as the interval of frequencies outside of which the spectral density is less than a definite specified amount. It turns out that Kotel'nikov's theorem in the general sense, that the function is defined over the interval T by

$$n = 2FT$$

"coordinates," is preserved when generalized in this manner.[1]

15. Fourier Integral and Discrete Spectra

The spectrum of a function f(t) is computed from the equation

$$S = \int_{-\infty}^{+\infty} f(t) e^{-j\omega t} dt. \tag{15.1}$$

This calculation is fulfilled with strict observation of the usual limitations imposed on a function f(t). One of these limitations, as we know, consists of the fact that the function f(t) should be absolutely integrable between infinite limits. It is evident that this condition is not fulfilled in the case when f(t) is a periodic function. Therefore, Eq. (15.1) is immediately inapplicable for computing the discrete (line) spectra of periodic functions.

It is possible, however, to formulate a certain generalization, which, even though it may not be mathematically rigorous, in many respects will be useful and of interest.

Let us first explain what is meant by a discrete spectrum from the mathematical point of view. Let us consider for simplicity a purely sinusoidal oscillation with frequency ω_0. The spectrum of such an oscillation is represented by a single spectral line with frequency $\omega = \omega_0$. The spectral density at this frequency is equal to infinity; at all the remaining values of the frequency it is equal to zero. In addition, from the definition of spectal density (see Section 3)

$$S = \pi \frac{dC}{d\omega},$$

where C is the amplitude, it follows that the integral of the spectral density over the frequency, where the integral is taken over any interval (ω_1, ω_2), including within its limits $\omega = \omega_0$, will be equal to

$$\frac{1}{\pi} \int_{\omega_1}^{\omega_2} S\, d\omega = \int_{\omega_1}^{\omega_2} \frac{dC}{d\omega} d\omega = C,$$

[1] We shall not treat this subject here more extensively, since it has the character of theoretical information and transcends the intended scope of our presentation. In this respect, see the report by A. N. Kolmogorov presented at the Session of the Academy of Sciences of the USSR on the Problems of Automation [in Russian] (Moscow, 1956).

i.e., the "area" of the contour traced out by a graph of the spectral density on a spectral diagram is a constant finite quantity equal to the amplitude of the analyzed sinusoidal oscillation.

However, the described properties of the function representing the spectral density exactly coincide with the properties of the unit pulse δ (see Section 11). Thus the spectral density of a sinusoidal oscillation with frequency ω_0 can be written in the form

$$S = \pi C \delta(\omega - \omega_0). \tag{15.2}$$

This concept is applicable to a constant component, for which it is necessary to assume $\omega_0 = 0$.

Therefore, the spectral density of any periodic process can be expressed by the equation

$$S = \pi \sum_{k=0}^{\infty} C_k \delta(\omega - k\omega_1),$$

where C_k is the amplitude of the kth harmonic, ω_1 is the fundamental frequency.

Let us turn now to the Fourier transform. If we apply the transformation (15.1) to the unit pulse, taken as a time function, then for its spectrum we obtain

$$S = \int_{-\infty}^{+\infty} \delta(t) e^{-j\omega t} dt = 1. \tag{15.3}$$

Substituting this result into the equation for the inverse transform:

$$f(t) = \frac{1}{2\pi} \int_{-\infty}^{+\infty} S(\omega) e^{j\omega t} d\omega, \tag{15.4}$$

we should obtain

$$\frac{1}{2\pi} \int_{-\infty}^{+\infty} e^{j\omega t} d\omega = \delta(t). \tag{15.5}$$

This relation, the validity of which is not at all obvious, can, however, be taken as a conditional integral definition for the unit pulse $\delta(t)$.

If we are agreed to adopt this definition, then, after modifying it, we can immediately write the further series of relations:

$$\frac{1}{2\pi}\int_{-\infty}^{+\infty} e^{j\omega(t-\tau)}\, d\omega = \delta\,(t-\tau), \tag{15.6}$$

$$\frac{1}{2\pi}\int_{-\infty}^{+\infty} e^{j\omega t}\, dt = \delta\,(\omega), \tag{15.7}$$

$$\frac{1}{2\pi}\int_{-\infty}^{+\infty} e^{j(\omega-\omega_0)t}\, dt = \delta\,(\omega-\omega_0). \tag{15.8}$$

Moreover, the signs of the exponents can be changed on the basis that δ is an even function.

The discussion which follows, if not serving as a proof of the validity of the relation (15.5), will nevertheless clarify its interpretation. This discussion is based on the fact that we choose some continuous function to which the Fourier transform is applicable, and from this function make a limiting transition to the unit pulse. A set of functions is most suitable for this purpose. We shall choose the bell pulse, i.e.,

$$f(t) = e^{-\beta^2 t^2} \tag{15.9}$$

and write

$$S\,(\omega) = \int_{-\infty}^{+\infty} e^{-\beta^2 t^2} e^{-j\omega t}\, dt = \frac{\sqrt{\pi}}{\beta}\, e^{-\frac{\omega^2}{4\beta^2}}. \tag{15.10}$$

Now let β tend to zero. Here the function $e^{-\beta^2 t^2}$ will approach unity, while the function

$$\frac{1}{2\pi}\, S = \frac{1}{2\sqrt{\pi}\beta}\, e^{-\frac{\omega^2}{4\beta}}$$

will approach the unit pulse. In fact, with increasing β this function will tend to zero for all values of $\omega \neq 0$; when $\omega = 0$ it will tend to infinity. However, the integral of this function will remain constant in value and equal to unity:

$$\frac{1}{2\sqrt{\pi}\beta}\int_{-\infty}^{+\infty} e^{-\frac{\omega^2}{4\beta^2}}\, d\omega = \frac{1}{\sqrt{\pi}}\int_{-\infty}^{+\infty} e^{-x^2}\, dx = 1.$$

Therefore, all the properties of the unit pulse are present, and it has been shown that as $\beta \to 0$, Eq. (15.10) passes to the limit (15.7), which is equivalent to (15.5).

In conclusion, we note that a harmonic oscillation and a unit pulse constitute a pair of conjugate functions, the relations between which are evident from the following table, in which the duplicity of the frequency-time representations already noted once again appear.

$f(t)$	$e^{j\omega_0 t}$	$\delta(t-\tau)$
$S(\omega)$	$2\pi\delta(\omega-\omega_0)$	$e^{-j\omega\tau}$

CHAPTER II

ANALYSIS

16. Posing the Problem

In setting out to present the problems of analysis we shall first of all attempt to define the subject under discussion. By analysis we mean the process of obtaining a spectrum. However, it must be emphasized from the start that we are not concerned with the analysis of functions, but of physical processes. Let us consider this in greater detail.

If the problem consisted of analyzing a function, i.e., finding the spectrum of a function, then this problem would be solved in another way, depending on how the function was specified. It can be specified in one of three ways: by its analytic expression, its graph, or by tables.

In the first instance the spectrum is computed analytically; in the other two instances either graphic-analytic numerical methods for computing the spectrum or special instruments—analyzers—of the mechanical, optical-mechanical, or optical-electrical type are employed. All these methods have been fully developed and described in every detail; the reader will find sufficient information on these problems in an encyclopedic work by Serebrennikov [14].

This entire area, however, is of no present interest to us. We are concerned with the problems of physical analysis. This form of analysis is characterized by the fact that the spectrum of some process is obtained during the course of a process as the result of its action on a definite physical instrument called the analyzer. Thus the problem of analysis remains a purely physical problem and, as we can see, its solution is connected with quite a number of purely physical peculiarities.

As already mentioned in Section 3, the analysis of the most multifarious physical effects can be demanded; we might be interested in the spectra of mechanical quantities—forces, velocities, accelerations, displacements, moments, etc.; electrical quantities—currents, potentials, charges, inductions, etc.; thermal, acoustic, and many other quantities. It would be extremely inconvenient to have to construct analyzers for each kind of quantity to be analyzed. This is not necessary in this day and age. The fact is that the modern approach in the area of measurement techniques is that all forms of measurements are conceptually reducible to electrical measurements. This tendency is justified, first of all, by the presence of a tremendous assortment of electrical instrumentation that is highly rated with respect to accuracy and is extremely sensitive, and, secondly, by the specific flexibility of electrical measurements. Without delving into the details, it suffices to note the possibility of setting up the instrumentation at any distance from the object of measurement, along with the fact that electrical measurements permit the measurement of rapidly changing quantities. Electromechanical oscillographs serve this purpose, and for particularly rapid processes electronic oscilloscopes are used.

The electrical methods that are very extensively applied in modern engineering for the measurement of nonelectrical quantities are based on the application of instruments which convert the measured quantity into one electrical quantity or another. These instruments are usually called pickups. Every apparatus for the electrical measurement of a nonelectric quantity basically consists of two elements: the pickup and the associated electrical instrumentation. Thus, e.g., the measurement of high temperatures is performed with an electrical pyrometer consisting of a thermocouple and millivoltmeter; the measurement of acoustic pressure is effected by a combination of high-quality microphone and vacuum-tube voltmeter having the proper amplification. In these two examples the thermocouple and the microphone are transducers and fulfill the function of pickups. There is in existence a tremendous quantity of the most diversified pickups, which make it possible to convert any quantity suitably subjected to measurement into an electrical current or potential with the possibility of realizing demands of the highest order as far as measurement is concerned.

With this introduction it becomes all the more evident that the analysis of any physical process consisting of the time variation of some physical quantity or other can be reduced to the analysis of an electrical process, i.e., correspondingly to a variable current or potential. Therefore, all modern engineering analyzers are electrical instruments.

We have defined analysis as the operation of determining a spectrum. A spectrum represents a set of amplitudes of the components of different frequencies. Consequently, an analyzer is an instrument that permits the measurement of the amplitude and frequency of each of the sinusoidal oscillations which go together to make up the complex oscillation to be analyzed.

Every analyzer is a measuring instrument. Therefore, we should devote considerable attention to the metrological characteristics of an analyzer and, first and foremost, to its accuracy. As we can see, errors in the analyzer are caused to a considerable extent by very special circumstances arising in the instrument during the process of analysis. In all that follows these circumstances will be discussed in as much detail as deemed necessary.

17. Spectral Instruments

For purposes of analysis we can avail ourselves of any instrument whose readings depend in one way or another on the frequency of the process affecting it. Any such instrument can be called a spectral instrument.

One of the following effects serves as the basis of operation of spectral instruments: interference, refraction in the presence of dispersion, resonance.[1]

The first two effects are used for constructing a whole series of spectral instruments in optics. Optical spectra are obtained by the interference method in interference spectroscopes, as well as in the diffraction grating. A spectrum can also be obtained by means of a prism when the index of refraction is grossly dependent on the frequency, or in other words, when dispersion of the phase velocity takes place.[2]

We shall not take too much time for the theory of these instruments; we are only interested in their general characteristics. The most essential feature is that the indicated spectral instruments are not only analyzers, but wave analyzers; they are especially equipped for the analysis of waves incident upon the analyzer. Therefore, they are most suitable in optics.

The resonator is an apparatus which makes use of the phenomenon of resonance for analysis, and is more universal, since it can be applied both for the analysis of waves, in which case it should be placed in a wave field, and for the analysis of concentrated actions. In optics, analysis by means of a resonator is impossible, for the simple reason that we are not yet in a state to construct an electrical resonator at frequencies of the order 10^{14} cps, with

[1] Moreover, for the purpose of analysis we can make use of the selectivity arising as a consequence of the orthogonality of the trigonometric functions. Section 20 is especially devoted to a discussion of this possibility.

[2] It should not be thought that the process of obtaining a spectrum on the basis of dispersion constitutes a monopoly in optics. An electrical communications cable also has dispersion; the higher frequency oscillations are propagated along it with a greater phase velocity. As a result of this at sufficiently great lengths (for example, transatlantic distances) of cable a spectral resolution of the signal <u>in time</u> is observed, an effect which is well known to long-distance engineers. This effect, however, insofar as the author knows, has not been applied for the analysis of measurements.

which we have to deal in optics.[3] The resonator has been found effective even at radio frequencies, let alone at the lower ultrasonic and acoustic frequencies. Since the analysis of various effects (except optical effects), as already mentioned, reduces to the analysis of the electrical current for the most part in circuits having lumped constants, the possible solutions to the problem of analysis are at the present time limited as follows: in optics, wave analyzers are applied exclusively, i.e., interference spectroscopes, diffraction gratings, and prismatic spectroscopes; for all the remaining effects occurring with frequencies from radio frequencies on down and permitting the transformation of the analyzed quantity into an electrical quantity (current or potential) the analysis is effected by means of resonators.

The simplest electrical resonator is the oscillation circuit consisting of a lumped inductance, capacitance, and resistance. However, this simpler form proves to be unsuitable in a number of cases.

First of all, the attenuation of an ordinary circuit is rather large, as a consequence of which the resolving power of the analyzer is small (this is discussed in greater detail in subsequent sections). The usual method for getting around this difficulty consists in replacing the electrical resonator by a mechanical type, which, as a rule, has considerably less damping. The circuit for the analyzer when applying a mechanical resonator becomes more complex: the mechanical resonator is included between two transducers. The first transducer converts the current into a mechanical force and excites the resonator. The second transducer reproduces the mechanical oscillation of the resonator and once again transforms this mechanical oscillation into an electrical oscillation. It is very convenient if the transducers and resonator are combined into a single element; thus we have piezoelectric and magnetostrictive resonators. The quartz plate of a piezoelectric resonator is simultaneously a transducer and vibrating system. The same applies to a magnetostrictive rod.

Secondly, it is frequently impossible in practice to construct a resonator at a specified frequency in the form of a system with lumped constants. In these instances one of the resonances of a system with distributed constants is used. But here it is interesting to note that the resonance in such a system—for example, in a section of wire or in a rod—is a wave effect, and here we arrive right back at the principle applied in optics. To be sure, the wave resonance in a section of transmission line is nothing more than the result of interference; a section of wire used as a spectral instrument is in no way differ-

[3]It might, however, be noted that there are natural resonators at optical frequencies—atoms and molecules. Such effects as, for example, resonance absorption, resonance fluorescence, etc., are caused by the action of these resonators.

ent from an interference spectroscope in the nature of the effects that take place within it. Thus the demarcation remarked above between the spectral instruments for optical and nonoptical effects is no longer as distinct. In order to complete the picture, it should be noted that the diffraction grating has been applied for the analysis of effects at ultrasonic frequencies. For this purpose the effect to be studied was converted into ultrasonic radiation in air (this operation is necessary, since the grating is a wave analyzer). The wave impinged on a diffraction grating of the appropriate dimensions. The diffraction spectrum was observed by means of an ultrasonic microphone. Such a complicated system hardly proves to be convenient. Later we shall devote special attention to analysis by means of resonators.

18. Simultaneous and Consecutive Analysis

The analysis of a complex oscillation by means of resonators can be effected in two different ways. The first method consists in applying an assembly of resonators tuned to different frequencies and simultaneously subjected to the action of the oscillation investigated. Such a method will be called simultaneous analysis. For the second method a single resonator with variable tuning is used. This form of analysis is called consecutive analysis.

It is evident that simultaneous analysis has an advantage over consecutive analysis in the speed of analysis. But, in addition, the two methods have a more significant difference. The fact is that the process of tuning the resonator in consecutive analysis requires a certain amount of time, and it is clear from the outset that—and we shall consider this in greater detail—the tuning cannot proceed too rapidly, otherwise the results of the analysis will be distorted by transient effects. Consequently, consecutive analysis is more suitable for periodic effects, or for effects whose character changes very slowly (i.e., varies little during the time of analysis). However, for the analysis of effects in which the behavior is rapidly changing, in particular, for the analysis of single pulses, consecutive analysis is quite unsuitable. It is, of course, true that very often we have the possibility of periodically repeating a pulse for the purpose of investigation. In this case consecutive analysis is applicable for the analysis of the pulses; recourse is made again and again to this possibility.

The essence of consecutive analysis is contained in the fact that the frequency of the resonator, varied smoothly, coincides successively with the frequencies of the harmonic components of the analyzed oscillation. The setting of the tuning device permits a reading of the frequency, while the amplitude of the harmonic is shown by an indicator fitted to the resonator. The common wavemeter can serve as an example of such an analyzer. Calibration of the wavemeter by means of a multivibrator is essentially the operation of analysis. It is true that in this case the analysis is qualitative, since the amplitudes of the harmonics are not of interest during calibration, but it is evident that they can be measured. The same wavemeter can be used successfully in analyzing, for example, a modulated oscillation. However, such a simple analyzer has not received extensive application for the following reasons.

Usually a very broad frequency range is demanded of a modern analyzer, whether it be an analyzer for mechanical, acoustic, ultrasonic, or radio frequencies. For example, the acoustic range encompasses about ten octaves. It is practically impossible to construct an adequately selective resonator with a smooth variation in tuning over such a range. Therefore, recourse is made to a modification of the method of consecutive analysis, a modification that is of great significance in the technique of analysis. This is contained in the fact that, instead of moving the resonance frequency along a frequency scale relative to fixed spectral lines, the whole spectrum is made to shift along the frequency scale relative to a fixed resonance frequency. In both this case and the former a successive coincidence of the individual spectral lines with the frequency of the resonator as a result of their relative motion on the frequency scale is observed.

In order to obtain a spectrum moving along the frequency scale it is necessary to make an appropriate transformation of the original spectrum. The required transformation is obtained quite simply: it is sufficient to multiply the analyzed oscillation by a sinusoidal potential with a variable frequency. This multiplication is carried out by some sort of modulator.

Let the analyzed oscillation be

$$x = \sum c_k \cos (k\omega_1 t + \varphi_k),$$

and let the auxiliary frequency (in German terminology, "Suchton" – literally, "searching tone") be

$$y = y_m \sin \Omega t,$$

where Ω is a frequency that is varied at will. Let us form the product

$$xy = y_m \sin \Omega t \sum c_k \cos (k\omega_1 t + \varphi_k) =$$

$$= \frac{1}{2} y_m \sum c_k \{ \cos [(\Omega - k\omega_1) t - \varphi_k] + \cos [(\Omega + k\omega_1) t + \varphi_k] \}.$$

We obtain the modulation spectrum, which reproduces the spectrum of the analyzed oscillation in the form of two side bands situated symmetrically relative to the carrier frequency Ω. It is now sufficient to choose the limits of variation of the auxiliary frequency Ω in such a way that one of the side bands of the converted spectrum passes completely through the resonator frequency. This modification of consecutive analysis has been used very extensively, fully justifying its important advantages. We shall not go into the engineering details at this point.

19. Static Resolving Power and Analyzer Error

The resolving power is an important instrumental characteristic of the analyzer. Resolving power is understood to mean in general its capability to resolve (or separate) two adjacent spectral lines. A quantitative measure of the resolving power is the smallest interval in frequency between two spectral lines for which they are still separated by the analyzer. The circumstances under which we can consider the lines separated must be clearly defined. In order to do this it is necessary to consider the process of analysis in more detail.

We shall begin with consecutive analysis. Let us assume that there is a single spectral line with frequency ω_1. Let the resonator be gradually tuned in such a way that its resonance frequency, after increasing, passes through the value ω_1. Here the indicator will give the highest reading. If now we record the reading of the indicator as a function of the resonance frequency, then this dependence will give nothing more than a resonance curve (Fig. 28).

Now let the analyzed spectrum consist of two spectral lines with the same intensity.

In this case at resonance there will exist simultaneously oscillations at two frequencies, and beats will arise with the difference in frequencies. The "amplitude" of the complex oscillation in this case will no longer be a constant quantity. However, we can agree that in speaking of the amplitude we shall mean its greatest value, which is equal to the sum of the amplitudes of both components of the oscillation. Each of these amplitudes will depend on the tuning of the resonator relative to the appropriate frequency, and consequently, by applying the foregoing discussion to each of the two spectral lines, we arrive at the conclusion that the readings of the indicator will give the graph of a double-humped curve, which is obtained upon compounding the two displaced resonance curves (Fig. 29). We see that a resonator with damping represents a separate spectral line of the resonance curve. (In general, it is impossible to construct an analyzer that would give a spectrum in the form of a line). But a spectrum consisting of many lines will be shown by the analyzer in the form of a smooth curve with a series of maxima corresponding to the individual spectral lines. The problem of analysis is thus solved, since the position of the maxima on the frequency scale will determine the position of the spectral lines, while the height of the maxima will determine the intensity of the lines. It now remains only to define the condition of resolution.

For the two resonance curves in Fig. 29 we can write

$$y_1 = \frac{1}{2} \frac{1}{\sqrt{\left(\frac{\omega - \omega_1}{\omega_1}\right)^2 + \frac{d^2}{4}}}, \quad y_2 = \frac{1}{2} \frac{1}{\sqrt{\left(\frac{\omega - \omega_2}{\omega_2}\right)^2 + \frac{d^2}{4}}},$$

where d is the attenuation of the resonator. Let us introduce

$$\omega_0 = \frac{\omega_1 + \omega_2}{2}, \quad b = \frac{\omega_2 - \omega_1}{2}.$$

Then

$$\omega - \omega_1 = \omega - \omega_0 + b = \Delta\omega + b, \quad \omega - \omega_2 = \omega - \omega_0 - b = \Delta\omega - b,$$

and the equation for the double-humped curve is

$$y = y_1 + y_2 \approx \frac{1}{2} \left[\frac{1}{\sqrt{\left(\frac{\Delta\omega + b}{\omega_0}\right)^2 + \frac{d^2}{4}}} + \frac{1}{\sqrt{\left(\frac{\Delta\omega - b}{\omega_0}\right)^2 + \frac{d^2}{4}}} \right].$$

When the damping is small the maxima of the curve will be situated at the frequencies ω_1 and ω_2, i.e., when $\Delta\omega = \pm b$; the value of the maximum is approximately equal to

$$y_{\max} \approx \frac{1}{d} + \frac{1}{2} \frac{1}{\sqrt{\left(\frac{2b}{\omega_0}\right)^2 + \frac{d^2}{4}}} \approx \frac{1}{d},$$

while the ordinate of the saddle, i.e., the magnitude of y when $\omega = \omega_0$, or $\Delta\omega = 0$, is equal to

$$y_0 = \frac{1}{\sqrt{\left(\frac{b}{\omega_0}\right)^2 + d^2}}.$$

The ratio of the ordinate of the saddle to the magnitude of the maximum is equal to

$$\frac{y_0}{y_{\max}} = \frac{d}{\sqrt{\left(\frac{b}{\omega_0}\right)^2 + \frac{d^2}{4}}} = \frac{2}{\sqrt{\left(\frac{2b}{\omega_0 d}\right)^2 + 1}}.$$

We see that this ratio depends on both the relative distance between the lines $\frac{b}{\omega_0}$ and the attenuation $\underline{d}$, where an increase in the separation is equivalent to a decrease in the attenuation, and vice versa.

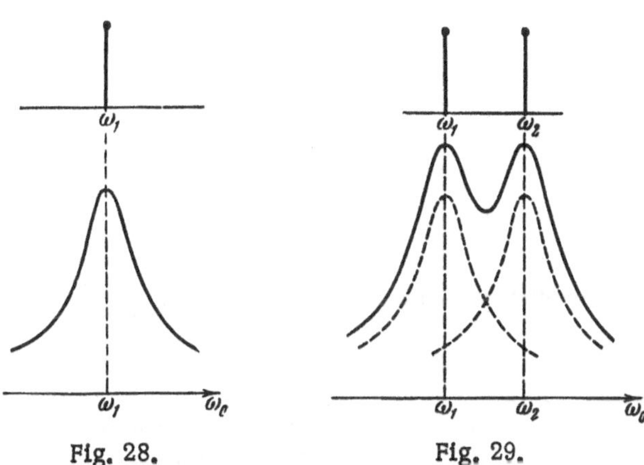

Fig. 28. Fig. 29.

Let us agree that two spectral lines of identical intensity are separable when the ratio $\frac{y_0}{y_{max}}$ does not exceed some definite value, for example $\frac{1}{2}$.

Then from the inequality

$$\frac{2}{\sqrt{1+\left(\frac{2b}{\omega_0 d}\right)^2}} \leqslant \frac{1}{2}$$

we find

$$\frac{2b}{\omega_0 d} \geqslant \sqrt{15} \approx 4$$

or

$$d \leqslant \frac{1}{4} \frac{2b}{\omega_0}.$$

Thus in the case considered the resolving power depends only on the attenuation of the resonator; the less the damping, the greater will be the resolving power; when the resolving power is specified the greatest permissible damping can be determined.

The relations derived are illustrated by the graph in Fig. 30, in which the double-humped curve is constructed for different values of the attenuation. As we can see, for a definite value of the attenuation the saddle flattens out, and the curve goes over to a single-humped curve. In this case it is obviously no longer possible to distinguish a separation in the lines. The magnitude of the attenuation corresponding to such an alteration in the form of the curve can be determined by setting $\frac{y_0}{y_{max}} = 1$; this gives

$$d = \frac{1}{\sqrt{3}} \frac{2b}{\omega_0}.$$

Let us now turn to simultaneous analysis. The analyzer consists of an assembly of resonators turned to different frequencies. All the resonators are simultaneously subjected to the oscillation being analyzed and react in some way to it. Let the analyzer consist of the resonators 1,2,3,4,..., whose resonance curves are illustrated in Fig. 31a. Let us further assume that the analyzed oscillation consists of a single spectral line, the position of which is marked by the vertical dashed line. Each of the resonators will respond to this oscillation in its own way; in Fig. 31a, the ordinates of each of the resonance curves are represented by small circles at those points where they give the values of the indicator readings for each resonator. The set of these readings, which from here on, for the sake of brevity we shall call the analyzer reading, is shown in Fig. 31b.

Thus, instead of the actually existing single line, the analyzer reading will be represented by a whole spectrum, the frequencies of which, in addition, depend on the tuning of the resonators, but not on the frequency of the oscillation being analyzed. However, any conclusion as to the complete impossibility of coping with simultaneous analysis would be too hasty. The fact is—and this is of the utmost importance—that the analyzer reading is single-valued, i.e., there is a fully definable connection between the spectrum of the analyzed oscillation and the analyzer reading. Let us consider this problem in somewhat more detail.

Let the analyzer be composed of a certain number of resonators differing in their tuning in such a way that the equation of the resonance curve for the kth resonator is written in the form

$$y_k = \frac{1}{2} \frac{1}{\sqrt{\left(\frac{\omega - \omega_k}{\omega_k}\right)^2 + \frac{d^2}{4}}},$$

where ω_k is the resonance frequency. Under the action of a sinusoidal oscillation with frequency ω_0 the analyzer reading is represented by the set of quantities

$$y_{k0} = \frac{1}{2} \frac{1}{\sqrt{\left(\frac{\omega_0 - \omega_k}{\omega_k}\right)^2 + \frac{d^2}{4}}}.$$

However, in view of the same approximations that were assumed in deriving the expression for the resonance curve,[1] the last relation can be rewritten in the form

$$y_{k0} = \frac{1}{2} \frac{1}{\sqrt{\left(\frac{\omega_k - \omega_0}{\omega_0}\right)^2 + \frac{d^2}{4}}},$$

where this expression means that the analyzer reading is represented by a set of lines with the frequencies ω_k, which are included in the same resonance curve, but with resonance at ω_0. The latter resonance curve is marked in Fig. 31b by a dashed line.

From this follows that if we put the analyzer together from a large number of resonators with resonance frequencies distributed uniformly and at small intervals over the frequency scale, then the reading of such an analyzer will give a dense enough spectrum that a determination of the position of the maximum becomes possible. On the basis of this assumption we can speak of the resolving power of the analyzer in the case of simultaneous analysis in exactly the same sense as in reference to consecutive analysis. It is also possible to consider the curve representing the reading of a consecutive analyzer as the

[1] The exact expression

$$y = \frac{1}{\sqrt{\left(1 - \frac{\omega^2}{\omega_k^2}\right)^2 + \frac{\omega^2}{\omega_k^2} d^2}}$$

for the case of a small frequency separation, i.e., for a selective resonator, becomes simplified, if it is assumed that

$$\omega \approx \omega_k, \quad 1 - \frac{\omega^2}{\omega_k^2} = \frac{(\omega + \omega_k)(\omega - \omega_k)}{\omega_k^2} \approx 2 \frac{\omega - \omega_k}{\omega_k},$$

which then leads to the approximate equation that we are using.

limiting case of the reading of a simultaneous analyzer when the number of resonators has been increased to infinity. Expressing this relation in still another way, it can be said that the readings for consecutive and simultaneous analysis are related in the same way as the spectra of singly occurring and periodically repeated pulses: the first spectrum is complex, the second is a line spectrum, which is enveloped by the first.

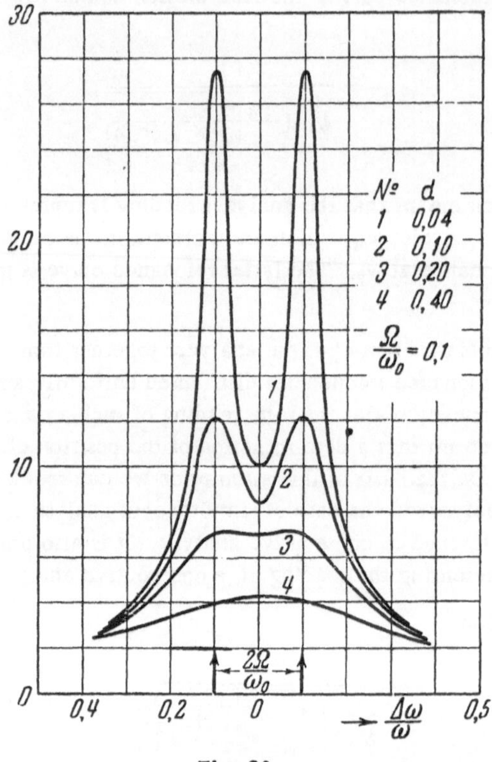

Fig. 30.

The possibility of good resolution by means of simultaneous analysis can be illustrated by reference to the well known reed frequency meter. This apparatus consists of an assembly of mechanical resonators with a large selectivity. If the tuning is carried out, as usual, in steps of 1 cps, then the frequency can be read with assurance to an accuracy of 0.5 cps, since when the frequency coincides with the middle of the interval between the frequencies of two adjacent reeds they oscillate simultaneously with equal amplitudes.

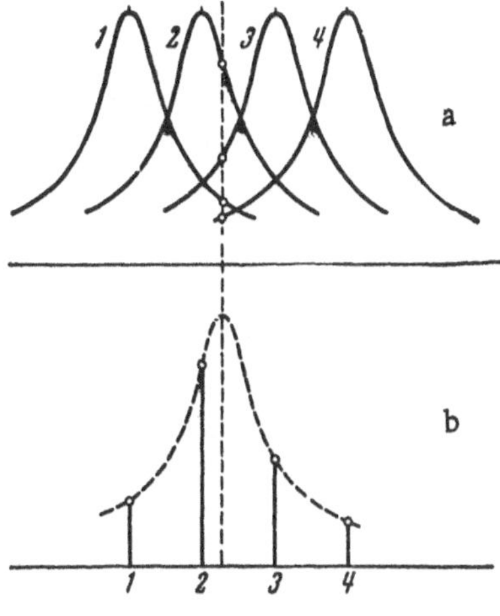

Fig. 31.

In principle it is possible to put the question of analysis in another way, so that it is particularly applicable to simultaneous analysis. Let us imagine that the analyzer is composed of an assembly, not of the usual resonators, but of ideal band filters with closely adjacent boundaries (Fig. 32a). Then in the presence of one spectral line (Fig. 32b) the analyzer reading will be unique: the only filter responding to the oscillation corresponding to this line will be the one within the limits of the pass band of which the excitation frequency lies. The analyzer reading is shown in the example of Fig. 32c. However, it is clear that the value of the frequency is not precisely determined; we can only assert that it is to be found somewhere within the limits of the given band.

It is also clear the measure of the resolving power for such a band analyzer is the bandwidth. If within the limits of one of the bands more than one line of the spectrum investigated is found, then the analyzer has not separated them; it only gives a total value for all the components whose frequencies lie within the limits of the given band. Under these circumstances it is natural that the band analyzer should be used to measure power, since the power is expressed in the band by the simple arithmetic sum of the powers of the individual components.

In concluding the present section let us consider the problem of the analyzer error. If the investigated oscillation is sinusoidal, then the amplitude and frequency of the oscillation can in principle be determined by means of the analyzer exactly. If, on the other hand, a complex oscillation is to be analyzed, then a certain amount of error that cannot be removed arises. This error is

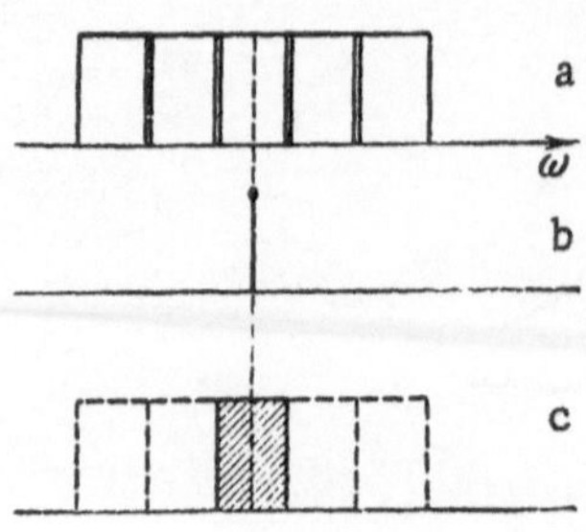

Fig. 32.

caused by the fact that all the components of the complex oscillation are acting simultaneously on the resonator; we are not in a position to separate their effect, and, as a consequence, we cannot determine the amplitude and frequency of each component individually.

Let the complex amplitude of the _k_th resonator under the action of the _i_th harmonic of the spectrum be

$$y_{ki} = \frac{x_i}{1 - \left(\frac{\omega_i}{\omega_k}\right)^2 + jd_k \frac{\omega_i}{\omega_k}},$$

where ω_i is the frequency of the harmonic, x_i is its amplitude, ω_k is the resonance frequency of the given resonator. Under the action of the entire spectrum the amplitude of the resonator will be

$$y_k = \sum_i \frac{x_i}{1 - \left(\frac{\omega_i}{\omega_k}\right)^2 + jd_k \frac{\omega_i}{\omega_k}}.$$

Let us assume that the resonator is tuned exactly at the frequency of one of the harmonics. Separating the components associated with this harmonic, we obtain

$$y_k = \frac{x_k}{jd_k} + \sum_{i \neq k} \frac{x_i}{1 - \left(\frac{\omega_i}{\omega_k}\right)^2 + jd_k \frac{\omega_i}{\omega_k}}.$$

The summation in this equation is carried out over all the remaining harmonics, i.e., over all _i_ that are not equal to _k_. Let us now assume that the spectrum is harmonic, i.e., that

$$\omega_i = \Omega i,$$

where Ω is the fundamental frequency. Moreover, let us assume that the relative attenuation for all the resonators is the same, i.e., let

$$d_k = d.$$

Then

$$y_k = \frac{x_k}{jd} + \sum_{i \neq k} \frac{x_i}{1 - \left(\frac{i\Omega}{\omega_k}\right)^2 + jd\,\frac{i\Omega}{\omega_k}}.$$

This expression shows that the reading from the resonator is proportional to the amplitude of the corresponding harmonic, i.e.,

$$|y_k| = \frac{x_k}{d}$$

only under the condition that the sum on the right-hand side can be neglected. Therefore, this sum then represents the error accumulated during analysis.

Considering the equation, we see that the error is less as the spectrum becomes sharper, i.e., the greater the fundamental frequency Ω. Moreover, the relative error falls off with diminishing attenuation (i.e., with increased resolving power), inasmuch as the first term on the right-hand side increases in comparison with the summation.

All of these considerations are equally applicable to both simultaneous and consecutive analysis.

20. Analysis Without Resonators

So far it has been assumed that analysis is carried out by means of a resonator, i.e., some apparatus having a definite frequency of selection. The simplest resonator is an oscillation circuit; a band filter can serve as a more highly selective instrument.

There is, however, still another possibility, which we shall consider in general terms. Let us draw the schematic shown in block form in Fig. 33. In this diagram M is a multiplier (ideal modulator), i.e., an apparatus at whose output a potential proportional to the product of the potentials delivered to the inputs of the following two elements is obtained: G, a sinusoidal oscillator with variable frequency and initial phase; I, an integrating circuit. If the generator supplies the potential

$$U = \cos \omega t,$$

then at the output we obtain

$$A_t(\omega) = \int_0^t f(\tau) \cos \omega\tau \, d\tau$$

(assuming that the circuit is switched in at the instant $t = 0$). If the initial phase of the generator potential is changed by $\pi/2$, then at the output we get

$$B_t(\omega) = \int_0^t f(\tau) \sin \omega\tau \, d\tau.$$

The quantities A_t and B_t are the cosine and sine components of the running spectrum. Connecting two circuits such as the one in Fig. 33 together with two generators giving sinusoidal potentials out of phase by $\pi/2$ and appending an element in which the square root of the sum ot the squares of A_t and B_t is formed, we obtain an analyzer at whose output the running amplitude spectrum of the function $f(t)$ is obtained:

$$\Phi_t(\omega) = \sqrt{A_t^2(\omega) + B_t^2(\omega)}.$$

The method described for obtaining the spectrum can be modified in such a way that it becomes possible to by-pass one of the circuits in Fig. 33. In order to do this it is necessary at a given frequency ω to vary the phase φ of the generator potential. Then at the output of the circuit in Fig. 33 we have

$$x(t,\omega,\varphi) = \int_0^t f(\tau) \sin(\omega\tau + \varphi) \, d\tau.$$

Let us find the greatest value of $\underline{x}$, considering the phase φ to be variable. For this we equate the derivative $\frac{\partial x}{\partial \varphi}$ to zero:

$$\begin{aligned} \frac{\partial x}{\partial \varphi} &= \int_0^t f(\tau) \cos(\omega\tau + \varphi) \, d\tau = \\ &= \cos\varphi \int_0^t f(\tau) \cos \omega\tau \, d\tau - \sin\varphi \int_0^t f(\tau) \sin \omega\tau \, d\tau = \\ &= A\cos\varphi - B\sin\varphi = 0. \end{aligned}$$

From this

$$\cos\varphi = \frac{B}{\sqrt{A^2 + B^2}}, \qquad \sin\varphi = \frac{A}{\sqrt{A^2 + B^2}}.$$

Substituting these values into the expression $\underline{x}$, we obtain

$$x_{\max}=\frac{A^2}{\sqrt{A^2+B^2}}+\frac{B^2}{\sqrt{A^2+B^2}}=\sqrt{A^2+B^2}=\Phi_t(\omega).$$

Thus, by choosing the phase so that a maximum is obtained (at a given frequency) in the output potential of the circuit in Fig. 33, then the amplitude spectrum (absolute value of the complex spectrum) is obtained directly.

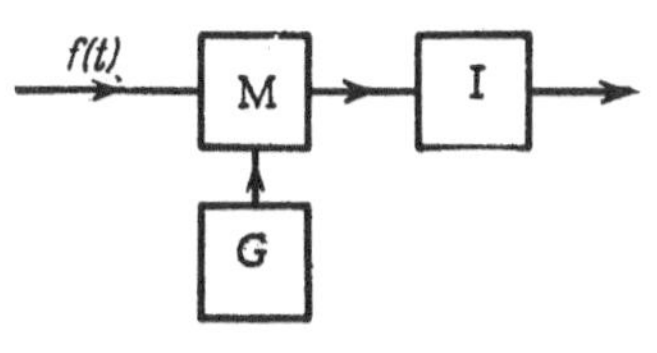

Fig. 33.

The circuit described above behaves in total correspondence with the mathematical definition of a spectrum, fulfilling all the operations introduced in this definition; in short, the circuit in Fig. 33 performs a Fourier transformation. Consequently, this circuit is the ideal analyzer, one that has an infinite resolving power.

However, now it must be noted that the circuit has the properties of an ideal analyzer only because its component parts live up to their designation in an ideal manner; we assumed that the modulator was an ideal multiplier, that the integrator integrated perfectly, and that the generator gave a sinusoidal potential without blemish. In a real circuit all of these conditions, of course, fail to be realized to some extent.

We shall now consider a circuit in which one of the ideal members is replaced by a real one, namely, the following: instead of the ideal integrator I let us connect a simplified integrating circuit in the form of an RC-element into the circuit (Fig. 34). Let us assume for simplicity that the potential to be analyzed is purely sinusoidal, and that we are considering the steady-state regime of the analyzer. Under steady-state conditions the transmission coefficient of the integrating circuit in Fig. 34 is equal to

$$K=\frac{\dot{U}_2}{\dot{U}_1}=\frac{1}{1+j\omega RC}.$$

The potential delivered to the input of the integrating circuit is equal to the product of the analyzed potential and the generator potential (let the initial phases of both potentials coincide):

$$U_1=\cos\Omega t\cos\omega t=\frac{1}{2}\cos(\omega+\Omega)t+\frac{1}{2}\cos(\omega-\Omega)t.$$

For the complex amplitudes of the potential at the output we can write

$$\frac{1}{2}\,\frac{1}{1+j(\omega+\Omega)RC}, \qquad \frac{1}{2}\,\frac{1}{1+j(\omega-\Omega)RC}.$$

When ω is not much different from Ω the first quantity can be neglected, and we obtain

$$|K| \simeq \frac{1}{2}\,\frac{1}{\sqrt{1+(\omega-\Omega)^2\tau^2}}$$

or

$$|K| \simeq \frac{1}{2}\,\frac{\frac{1}{\Omega\tau}}{\sqrt{\left(1-\frac{\omega}{\Omega}\right)^2+\frac{1}{(\Omega\tau)^2}}},$$

where

$$\tau = RC$$

is the time constant of the integrating circuit.

The expression obtained differs only by a constant factor of $1/\Omega\tau$ from the expression for the transmission coefficient of a simple oscillation circuit:

$$|K| = \frac{1}{\sqrt{\left(1-\frac{\omega^2}{\omega_0^2}\right)^2+\frac{\omega^2}{\omega_0^2}d^2}} \simeq \frac{1}{2}\,\frac{1}{\sqrt{\left(1-\frac{\omega}{\omega_0}\right)^2+\frac{d^2}{4}}}.$$

It thus turns out that the analyzer circuit we are considering has exactly the same properties with respect to selectivity as the ordinary resonator, so that, consequently, all that was said earlier about the resolving power and error of a resonator analyzer is applicable.

It is interesting that in the analyzer considered, the role of the attenuation is taken by the quantity $1/\Omega\tau$. Since for increasing the time constant τ = RC there are no practical limitations, it is possible to have an analyzer with a very high resolving power. On the other hand, the resolving power can easily be varied within wide limits by varying R or C. Making use of this fact, it is possible, for example, to investigate the fine details of any part of the spectrum.

The technical production of the system described is possible in the most diversified ways. As an instructive example we shall cite the ordinary wattmeter, which is subjected to considerable application as an analyzer in recent times. For example, if we wish to apply an electrodynamic instrument as an analyzer, it is sufficient to let the analyzed potential and the generator potential be the corresponding potentials on the two windings of the motor (stator and rotor). The rotational moment is porportional to the product of the currents in the windings; thus modulation is effected. Integration is carried out in the mechanical part of the apparatus (as a result of the ballistic properties of the rotor). If we wish to multiply the potentials directly (rather than the currents that are proportional to them), then, instead of an electrodynamic wattmeter, we may use a quadrant electrometer. The readings in instruments of this kind are usually not taken from the constant deflection (with a null in the frequency difference), but from the amplitude of the oscillations of the moving part of the instrument at the low frequency ω_0 of its mechanical frequency. We shall describe the effects that occur here.

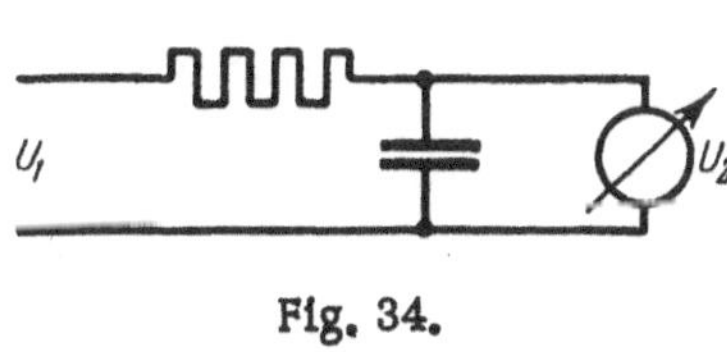

Fig. 34.

If the auxiliary frequency ω varies slowly, passing through the value Ω, then the variable frequency difference $|\Omega - \omega|$ will act on the instrument, first diminishing to zero, then increasing again. In this case the moving system of the apparatus oscillates, the amplitude of the oscillation has two maxima, when the difference in frequencies is equated with the natural frequency of the moving system. The process taking place in the measuring instrument with a slow variation in ω is shown in Fig. 35. The amplitude is at first small, then increases to a maximum, decreases to some constant deflection, and then everything proceeds in the opposite order. The envelope of the process turns out to be nothing other than the resonance curve of the moving system of the instrument. From the point of view of the fundamental principle of an analyzer we should have taken the reading at $\omega = \Omega$, where the arrow stops. But in practice it is very much more convenient to take the reading at points where the moving system resonates.

In conclusion, it is necessary to give an answer to the question that naturally comes up: does any principal difference exist between analysis by means of resonators and analysis by the method described above?

There is, of course, a difference, and it is of theoretical significance. This is the same difference that exists between filtering and heterodyning. The method described in this section is essentially a method of frequency transformation, or in particular, a method of synchronous detection with

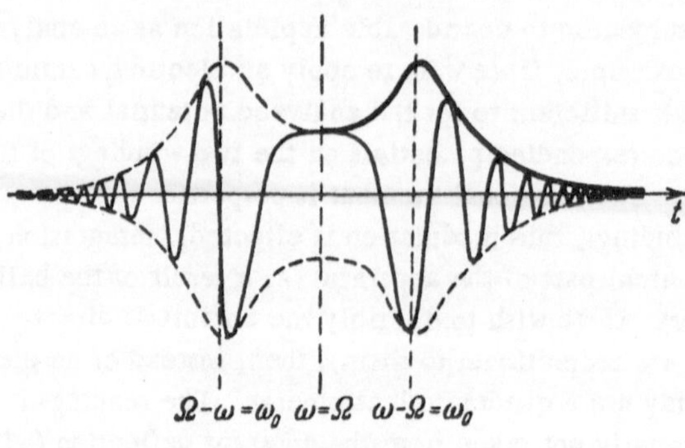

Fig. 35.

respect to some component of the spectrum. The difference in principle is contained in the fact that filtering is realized by means of a passive line system with constant parameters, whereas the heterodyning analyzer considered from the physical-mathematical point of view reduces to a linear system with variable parameters.

21. Resonator Operation

The action of an analyzer with resonators has been taken for granted so far, it being understood that when the resonator has frequency selectivity it is able to separate certain spectral components, i.e., to perform analysis. However, in light of the discussions of the preceding section the question naturally arises: is the resonator actually capable of analysis, or, in other words, what properties must the resonator have in order that the quantity observed at the output actually reflects the spectrum of the function analyzed? In order to give an answer to this question we must calculate under what conditions the quantity at the output of the resonator is connected with the input quantity through a Fourier transform. If such a connection exists, then this means that the resonator must fulfill its designated function.

In the general case, when the arbitrary function $f(t)$ is fed to the input of the resonator at the instant $t = 0$, the resonator's response can be expressed by the Duhamel integral

$$x(t) = \int_0^t f(\tau)\, g(t-\tau)\, d\tau, \tag{21.1}$$

where $g(t)$ is the time characteristic (pulse reaction) of the resonator. Let us assume that

$$g(t) = \sin \omega_0 t. \tag{21.2}$$

In this case

$$x(t) = \int_0^t f(\tau) \sin \omega_0 (t - \tau)\, d\tau =$$

$$= \sin \omega_0 t \int_0^t f(\tau) \cos \omega_0 \tau\, d\tau - \cos \omega_0 t \int_0^t f(\tau) \sin \omega_0 \tau\, d\tau =$$

$$= A \sin \omega_0 t - B \cos \omega_0 t.$$

Here A and B are, as before, the cosine and sine components of the running spectrum. They are time functions, and can be considered as slowly varying components of the amplitude of the oscillation x(t). The envelope (i.e., the time-dependent amplitude) of this oscillation is obviously equal to

$$\sqrt{A^2 + B^2} = \Phi_t(\omega_0),$$

i.e., to the running amplitude spectrum. Thus, if for the purpose of analysis a resonator with the time characteristic (21.2) is applied, then the analysis is carried out in correspondence with the mathematical definition of a spectrum, i.e., we have an ideal analyzer. But the time characteristic (21.2) is ascribed to a lossless loop, i.e., a circuit consisting of L and C (R = 0). Therefore, the next question that arises is whether the real resonator really serves as an analyzer. For a loop with losses

$$g(t) = e^{-\alpha t} \sin \omega_0 t$$

and

$$x(t) = \int_0^t f(\tau)\, e^{\alpha(\tau - t)} \cos \omega_0 \tau\, d\tau =$$

$$= \sin \omega_0 t \int_0^t f(\tau) e^{\alpha(\tau - t)} \cos \omega_0 \tau\, d\tau - \cos \omega_0 t \int_0^t f(\tau)\, e^{\alpha(\tau - t)} \sin \omega_0 \tau\, d\tau =$$

$$= A_\alpha \sin \omega_0 t - B_\alpha \cos \omega_0 t.$$

In this expression A_α and B_α are the cosine and sine components, respectively, of the running spectrum, but not the function $f(\tau)$; instead, the weighted function

$$f_\alpha(\tau) = e^{\alpha(\tau - t)} f(\tau).$$

The sliding weighted function appearing here is the same function that was used by Fano (see Section 6); the above considerations thus give basis to the definitions assumed by Fano.

In the more general case of a narrow-band resonator of arbitrary form the time characteristic can be written in the form

$$g(t) = p(t) \sin \omega_0 t + q(t) \cos \omega_0 t = \operatorname{Re} s(t)\, e^{j\omega_0 t},$$

where ω_0 is the center frequency of the band transmitted by the resonator. In this case

$$x(t) = \int_0^t f(\tau)\, g(t - \tau)\, d\tau = \int_0^t d\tau\, f(\tau) \operatorname{Re} s(t - \tau)\, e^{j\omega_0 (t-\tau)} =$$

$$= \operatorname{Re} e^{j\omega_0 t} \int_0^t f(\tau)\, r(\tau - t)\, e^{-j\omega_0 \tau}\, d\tau$$

and the integral represents a complex running spectrum with the weighted function

$$r(-t) = s(t) = p(t) - jq(t).$$

The final result following from all that has been said is contained in the fact that every real resonator does not give us the true spectrum of the analyzed function, but the spectrum of a weighted function, where the weighting function depends on the characteristics of the resonator. We obtain the true spectrum only in the case when the time characteristic of the resonator approaches

$$g(t) = \sin \omega_0 t,$$

while the weighting function, as a consequence, goes to unity. On the basis of these considerations the error of a real analyzer may be estimated. However, for the moment we shall not concern ourselves with this (see Section 26).

From the discussion it is clear that the metrological characteristic of an analyzer should contain data as to the character of the resonator, or at least as to the time constant. Otherwise it is impossible to establish what exactly is being measured by the analyzer. Unfortunately, in the majority of cases only the transmission bandwidth of the resonator is indicated, which is not enough.

As an illustration, let us refer to the interesting application of weighted spectra, treated as instantaneous spectra, in particular, the so-called "visual" speech [28]. The essence of the matter is contained in the fact that an instantaneous spectrum (weighted by means of band filters) is shown directly on a moving screen. Since the instantaneous spectrum is a function of two variables—frequency and time—the spectrum is shown on the plane of the screen in rectangular coordinates; time is plotted along the horizontal, frequency along the vertical. As for the function of these two variables, i.e., the spectral density, it is reflected in the brightness of the screen illumination at the given point (and in recording permanently by the degree of darkening of a special photographic paper). Speech is analyzed in a range extending approximately to 4 kc, which is divided into bands of 300 kc in width each. The pattern obtained gives a sufficiently complete representation of the sounds of speech; it can be instructed to read "visual speech" directly from the screen. Orignally the "visual speech" apparatus was designed for the deaf—in order to replace audible perception by visible— and, in fact, was successfully used for this purpose. However, this kind of technique can also be used for correcting speech defects, for phonetic investigations, as well as for automatic machines capable of fixing the sounds of speech or of carrying out commands sent by the voice, etc.

22. Actual Operating Conditions of the Analyzer

The conclusions of Section 19 need serious corrections in view of the actual operating conditions of an analyzer, and we shall proceed immediately to a development of these.

Right in the beginning it was noted that a periodic function of the form

$$f(t) = f(t + nT)$$

is a mathematical abstraction. But even if such an effect, corresponding to this definition, were to exist, then it would still be nonexistent for the analyzer. For the analyzer is in reality turned on at a finite time and, consequently, the effect analyzed represents for the analyzer a segment over the period of time from the instant of switching on to the instant of switching off.

The analyzer can give the true spectrum of an effect in the one existing case when the analyzed effect is entirely contained within this segment of time, i.e., when a known nonperiodic effect of the pulse type, which is equal to zero outside a finite interval of time is analyzed, and under the conditions that this interval of time, i.e., the pulse duration, be less than the time during which the analyzer is switched on. During the analysis of extended effects, however, periodic or nonperiodic, the analyzer is in principle not in a condition to give the true spectrum of these effects. By definition a spectrum is expressed by the integral

$$S = \int_{-\infty}^{+\infty} f(t)\, e^{-j\omega t}\, dt. \tag{22.1}$$

If the analyzer fulfills the required multiplication and integration, then in no event can it perform the integration within infinite limits. The analyzer "does not know" what has gone on before it was switched on, and likewise "does not know" what will go on after it is switched off. It can only perform the integration over the interval $t_1 < t < t_2$, where t_1 and t_2 are the instants of switching on and switching off, respectively.

The best that can be expected of the analyzer, operating continuously, is that the integration will be performed over the interval from the instant of switching on until the immediate moment in time. If the instant of switching on is taken as the time origin, then the analyzer can measure the quantity

$$S_t = \int_{0}^{t} f(t)\, e^{-j\omega t}\, dt, \tag{22.2}$$

i.e., the running spectrum of the function f(t) (see Section 5). The greater the time during which the analyzer is switched on, the more the running spectrum will approach the true spectrum of a prolonged effect. If the time during which the analyzer is on is very small, then whatever the nature of the effect is, the analyzer will only obtain a short pulse, the complex spectrum of which has a uniform density up to frequencies determined by the duration of operation, and the higher these frequencies, the less will the operating time have been, i.e., the shorter the pulse.

These considerations can easily be verified experimentally. A sinusoidal oscillation is perceived by the ear as a pure musical tone. If this tone is sounded over shorter and shorter intervals of time, then a distinct sensation of the pitch of the tone will be gradually lost, until, finally, a sufficiently short-lived sounding of the tone will be detected simply as a click, corresponding to a short pulse with a broad complex spectrum.

From all that has been said it follows that the result of analysis will depend in principle on the time. Allowance for this dependence can be made from two dissimilar (but leading, of course, to identical results) points of view: the time and spectral (frequency) concepts.

We shall give a detailed discussion of the process of analysis from both points of view.

a. Time point of view. We previously have proceeded to define the resolving power of a resonator from its resonance curve (frequency characteristic). But the resonance curve represents the behavior of the resonator in the steady-state regime. Steady-state is the name given to that regime in which the transient effects that arise when the resonator is turned on have ceased to exist, i.e., after an infinitely long time. In practice transient effects can be neglected after the passage of a short period of time. However, the settling time is greater as the selectivity of the resonator becomes greater, i.e., the less its attenuation. In application to analysis we are interested precisely in resonators with a high selectivity. Therefore, in investigating the operation of the analyzer, we cannot, generally speaking, ignore transient effects. Setting up and solving the corresponding differential equations, we obtain the amplitude of the resonator oscillations as a function not only of the frequency (as in the steady-state regime), but also of the time. Consequently, the resonance curve of the resonator will be different for different instants of time; it is properly called the dynamic resonance curve in contradistinction to the ordinary resonance curve, which we should call the static resonance curve and which is the limiting form of the dynamic resonance curve for $t \to \infty$.

Thus all the relations derived earlier retain their meaning only as limiting expressions. The true resolving power of the analyzer is determined by its dynamic resonance curve. Its maximum will become blunter, the shorter the time of operation; consequently, the analysis is more accurate, the longer the time of analysis.

b. Spectral point of view. We have also proceeded earlier to define the resolving power of the analyzer from the idea that a periodic, in particular, a sinusoidal oscillation was acting on the analyzer. In reality this is not the case: the analyzer is on only during some finite time. Consequently, for each given instant of time the action on the resonator can be represented by a nonperiodic effect, namely, a segment of a sinusoid, cut out from the instant that the analyzer is turned on until the immediate moment in time. The analyzer will analyze this effect with its intrinsic static resolving power. But since the spectrum of the action itself varies with the continuance of time ("running spectrum"), the analyzer reading will also vary. In particular, in the very first instants of analysis, when the periodic character of the effect is still undefined and the running spectrum is practically uniform, the analyzer will give almost identical readings at all frequencies. Only in the limit, when the complex running spectrum degenerates into a discrete spectral line, will the analyzer give the spectrum defined on the basis of the stationary representations presented earlier. These representation thus remain in effect in deriving the limiting relations. In reality, however, the analyzer reading turns out to be more diffuse, i.e., the resolving power is less, as the operating time of the analyzer becomes shorter.

Both points of view—time and spectral—are theoretically equivalent. From here on the time point of view will be adopted on the basis of the following methodological considerations:

1. The time point of view is more easily comprehensible and less formal in nature.

2. In adopting the time point of view, we can characterize the analyzer as such independently from the conditions under which it is turned on and the character of the effect to be analyzed.[1]

3. Certain issues are in general more difficult (although in principle it is possible) to outline in the language of spectra. In particular, attention is focused here on consecutive analysis. All that has gone before referred to simultaneous analysis, when we are concerned with the response to some kind of action of a resonator with fixed tuning. But if we turn to an analyzer containing a resonator with variable tuning, then we have to deal with more complex effects, and it proves to be more suitable to speak of these in the language of the time point of view.

In the case of consecutive analysis it is further evident that the properties of the resonator can be described by the static resonance curve only when the tuning is infinitely slow. In actuality, the analysis proceeds with a finite speed. It is easy to comprehend that, the more rapidly the tuning of the resonator passes through the frequency of a given spectral component, the lesser will be the degree to which this component can excite oscillation in the resonator. The same applies to the modified version with the application of an auxiliary frequency, in which case the tuning of the resonator is fixed, while the spectrum

[1]We are here concerned with the fact that in the analysis of a periodic effect we can take the finiteness of the operating time of the analyzer into account in two ways: 1) by characterizing the effect by an ordinary spectrum, while the analyzer is characterized by the dynamic frequency characteristic; 2) by characterizing the effect by its running spectrum, while the analyzer is characterized by the ordinary static frequency characteristic. The first method, which makes no attempt at a description of the analyzed effect, but, on the other hand, encompasses both the properties of the analyzer and its operating conditions through the dynamic characteristic, is the more natural and convenient.

of the investigated oscillation is shifted at some velocity or other along the frequency scale.

In the case of consecutive analysis the properties of the analyzer also can be conveniently described through the dynamic resonance curve. Theory and experiment both show that its maximum becomes blunter than in the case of the static resonance curve, where the blunter it becomes, the greater will be the speed of the relative motion along the frequency scale of the resonator tuning and the investigated spectrum, i.e., the greater the rate of analysis.

From all that has been said it follows that the concept of resolving power of the analyzer is deprived of any practical meaning, unless it is connected in one way or another with the rate (or duration) of analysis. It is evident that for the total characteristic of the properties of the analyzer we should introduce the concept of the dynamic resolving power of the analyzer, which would depend both on the parameters of the analyzer and on the speed of analysis. The subsequent sections will deal with the detailed investigation of this problem, where simultaneous and consecutive analysis will be considered separately, since the problem is set up in these two cases in essentially very different ways.

23. Relation Between Resolving Power of the Analyzer and Analysis Time

In the analysis of periodic effects a very general assumption holds, namely, that the greater the resolving power of the analyzer the greater will be the time necessary for analysis.

This assumption is easily explained in the example of the resonator. On the one hand, the resolving power of the resonator is greater as the attenuation becomes greater, i.e., the sharper the resonance curve of the resonator. On the other hand, the less the attenuation of the resonator, the more slowly will the natural oscillations that arise when the resonator is turned on die out, and, consequently, the longer it will be necessary to wait before the resonator can be considered in the steady-state regime. It is obvious that a reading from the resonator during the analysis of a periodic effect should be made in the steady state regime. Otherwise, it will be impossible to apply the static resonance curve of the resonator in defining its resolving power (more was said in detail on this matter in Section 22).

The example with the resonator shows that the assumption first stated refers to the influence of transient (nonsteady-state) effects in the analyzer, and in the cited formulation refers only to the analysis of periodic effects.

This assumption has, as stated, a very general character. We can cite still another example, in which the validity of this assumption does not appear obvious at first glance. This is in connection with the diffraction grating. The resolving power of a diffraction grating as an analyzer depends, as we know, not on the grating period, but on its overall dimension (on the number of lines). But when the size of the grating is increased, the path traversed by the waves and, consequently, the corresponding traversal time are increased. It is understandable that in optics this is of no practical significance, but we are interested in the theoretical side of the issue.

Although the essence of the matter in general is clear, we cannot, of course, be satisfied with an explanation of an assumption as to the connection between the resolving power and analysis time on the basis of isolated examples, despite their convincing nature. Our problem consists in proving this assumption in general form.

Let us imagine the selective element of the analyzer in the form of a quadrupole, to the input of which the quantity x is delivered, while at the output the quantity y is obtained, where the selectivity is manifested in the fact that the ratio

$$K = \frac{y}{x},$$

i.e., the transmission coefficient depends in some definite way on the frequency, when x and y are oscillatory quantities.

On the other hand, if the quadrupole is excited by an impulse at the input, then at the output a transient effect of some duration will arise. We have to find the connection between the width of the frequency characteristic (K) and the duration of the transient effect. The first quantity determines the resolving power, the second determines the time necessary for analysis. The relationship of interest to us has already been established in Section 12. To wit, it was shown that

$$K(\omega) = \int_{-\infty}^{+\infty} g(t)\, e^{-j\omega t}\, dt,$$

i.e., that the complex transmission coefficient $K(\omega)$ is nothing other than the spectrum of the function $g(t)$, which represents the effect at the output when an excitation of the form $\delta(t)$ is delivered to the input, and which we called the time characteristic.

We note, in passing, that in practice it is more suitable for us to use the function g(t) = h'(t), rather than the transient function h(t) [which corresponds to an excitation of the form σ (t)], for the purpose of defining the duration of the transient effects, since the latter function frequently tends to a constant limit with increasing t.

Thus the issue reduces now to establishing the connection between the duration Δt of the function g(t) and the width Δf of its spectrum [K(ω)]. This problem was considered in detail in Section 11, where it was established that

$$\Delta f \Delta t \geqslant \mu.$$

In application to the problem under consideration this relation is conveniently written in the form

$$\Delta f\, \Delta t = A \qquad (23.1)$$

and is interpreted in the following way: for a given system the time necessary for analysis is inversely proportional to the resolving power. This assumption, therefore, is proved. It still remains to indicate in sufficient detail just what is meant by the expression "given system."

The relation (23.1) would not have been of any interest to us, had it not implied some alteration in the properties of the system. The question is put as follows: if the function

$$f(a, b, c, \ldots, t),$$

where a,b,c,... are parameters, is given, and if, to be sure, the spectrum of this function is to depend on the same parameters, i.e.,

$$\Phi = \Phi(a, b, c, \ldots, \omega),$$

then it it permissible to assert that the constant A in Eq. (23.1) retains its import with any changes in the parameters. In such general form this assertion would be invalid. But we note that we are dealing, in particular, with the kind of functions in the expressions for which the parameters appear as multipliers associated with the independent variable. It is easily shown that if we multiply the argument by a constant quantity, then we have

$$f(\alpha t) = f(t_1), \quad \Phi\left(\frac{\omega}{\alpha}\right) = \Phi(\omega_1),$$

and, consequently, in this case

$$\Delta f\, \Delta t = \Delta f_1\, \Delta t_1 = A,$$

i.e., the product $\Delta f \Delta t$ does not change its magnitude when the parameters are varied.

Consequently, the given system in the formulation outlined above should be understood to mean a system corresponding to the given scheme or, speaking in more general terms, a system described by a differential equation of the given form, but without the assignment of special values to the coefficients of this equation.

Thus, e.g., if we decrease the attenuation of the resonator, then the analysis time will increase in exactly the same proportion as the resolving power.

We now note that one can arrive at Eq. (23.1) from another point of view. If A is a constant quantity for a given system, then how is the system chosen so as to get the least value of A? There certainly is no basis for our considering an ordinary resonator to be the optimum form of selective element for the purpose of analysis. Is it impossible to construct a system having better properties than the ordinary resonator, for example, having a shorter settling time, given the same resolving power, or a larger resolving power, given the same settling time?

From the discussion in Section 12 relevant to the bell pulse we can conclude that this form of pulse has particularly suitable properties in the sense of making the product $\Delta f \Delta t$ small. In order to be able to apply these properties for the purpose of analysis it is necessary to find some function that, preserving the bell character, will have a maximum at any specified frequency ω_0. It is easily imagined that if the spectrum of the ordinary bell pulse

$$f(t) = e^{-\frac{\pi}{2} t^2}$$

has the form

$$\Phi(\omega) = \sqrt{2}\, e^{-\frac{\omega^2}{2\pi}},$$

then the required property will be ascribed to the function

$$\Phi_1(\omega) = \sqrt{2}\, e^{-\frac{(\omega-\omega\ )^2}{2\pi}}.$$

On the basis of theorem (4.6) such a spectrum corresponds to the time function

$$f_1(t) = e^{j\omega_0 t} e^{-\frac{\pi}{2} t^2}.$$

This oscillation, which has a bell-curve envelope is shown in Fig. 36.

This kind of pulse is of great interest in the theory of communications, where it is considered the most advantageous elementary signal. For our purpose this function has another sense: it represents the time characteristic of some system which we hope to apply as the selective element of an analyzer. In other words, this function represents the effect at the output of a system upon excitation at the input in the form δ (t).

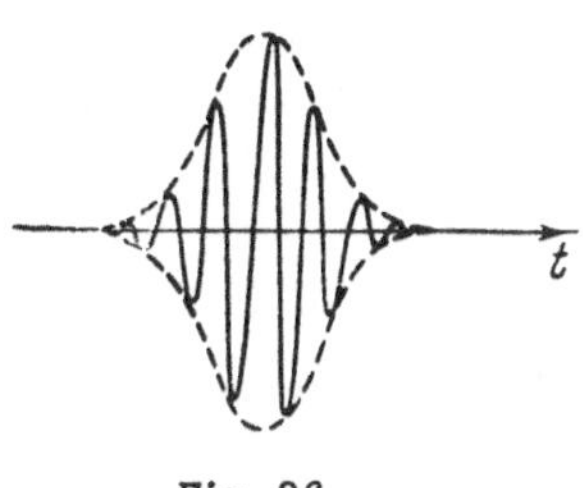

Fig. 36.

Certain general considerations can be expressed with respect to the properties of such a system. Examining Fig. 36, we note that the desired system possesses the essential property that oscillations are not stimulated immediately in it at the instant of striking, but gradually evolve until a maximum is reached, then begin to die off. Consequently, the first approximation to the desired system could be a combination of two weakly coupled resonators, of which the first received the impulse, then gradually delivered energy to the second resonator, at the output of which it would be possible in this way to observe a pattern similar in some degree to Fig. 36. It can be assumed that for such a combination the product $\Delta f \Delta t$ will prove to be less than for the ordinary resonator. We can also count on obtaining a better approximation to the required properties by adding more and more to the complexity of the system. And, in fact, the required characteristic belongs to a system consisting of a sufficiently large number of weakly coupled contours, which is brought about in practice in the form of a multi-stage resonance amplifier.

24. Dynamic Resolving Power of a Resonator

Let a resonator with variable tuning be turned on at t = 0 under the action of a sinusoidal excitation with frequency ω. Let us determine the dynamic characteristic of the resonator. We shall assume that the function of the resonator is fulfilled by an electrical circuit consisting of R, L, and C in series. The resonator equation has the form

$$L\frac{di}{dt}+Ri+\frac{1}{C}\int i\,dt=\sin\omega t \qquad (t>0)$$

or, differentiating with respect to t and letting L = 1,

$$\frac{d^2i}{dt^2}+2\alpha\frac{di}{dt}+\omega_0^2 i=\omega\cos\omega t \qquad (t>0). \tag{24.1}$$

We do not need the complete solution to this equation, only the value of the amplitude, which depends on the time, i.e., the expression for the envelope. In order to determine it we shall employ a method consisting in finding the solution in the form

$$i = a(t) \sin \omega t + b(t) \cos \omega t, \tag{24.2}$$

where a and b are slowly changing time functions, i.e., functions that vary slowly in comparison with sin ωt or cos ωt. The criterion for slowness is defined by the fact that the derivatives of these functions are small quantities of the first order in comparison with the derivative of sin ωt or cos ωt, i.e., in comparison with ω.

We shall further assume that α is also a small quantity of the first order, i.e., that we are dealing with a circuit of high selectivity. Differentiating (24.2), substituting into (24.1), and equating the coefficients separately for sin ωt and cos ωt, we obtain, after rejecting second-order small terms ($\ddot{a}$, $\ddot{b}$, $\alpha\dot{a}$, $\alpha\dot{b}$), the so-called abridged equations:

$$\left.\begin{aligned} (\omega_0^2 - \omega^2) a - 2\omega (\dot{b} + \alpha b) &= 0, \\ (\omega_0^2 - \omega^2) b + 2\omega (\dot{a} + \alpha a) &= \omega. \end{aligned}\right\} \tag{24.3}$$

These equations are now first-order equations, wherein the advantage of the method is contained. For the solution of Eqs. (24.3) we shall make use of an operational method. Let us write the operational representation of these equations:

$$\begin{aligned} (\omega_0^2 - \omega^2) \bar{a} - 2\omega (p + \alpha) \bar{b} &= 0, \\ (\omega_0^2 - \omega^2) \bar{b} + 2\omega (p + \alpha) \bar{a} &= \omega, \end{aligned}$$

whence

$$\bar{a} = \frac{1}{2} \frac{p + \alpha}{(p + \alpha)^2 + \beta^2}, \qquad \bar{b} = \frac{1}{2} \frac{\beta}{(p + \alpha)^2 + \beta^2},$$

where

$$\beta = \frac{\omega_0^2 - \omega^2}{2\omega}.$$

Applying the well known equations of operational calculus, we find

$$a(t)=\frac{1}{2}\frac{\alpha}{\alpha^2+\beta^2}\left[1-e^{-\alpha t}\left(\cos\beta t-\frac{\beta}{\alpha}\sin\beta t\right)\right],$$
$$b(t)=\frac{1}{2}\frac{\beta}{\alpha^2+\beta^2}\left[1-e^{-\alpha t}\left(\cos\beta t+\frac{\alpha}{\beta}\sin\beta t\right)\right].$$

We are interested in the envelope, which is expressed in terms of a and b as follows:

$$c(t)=\sqrt{a^2(t)+b^2(t)}.$$

Carrying out the computations, we find

$$c=\frac{1}{2}\sqrt{\frac{(1-e^{-\alpha t}\cos\beta t)^2+(e^{-\alpha t}\sin\beta t)^2}{\alpha^2+\beta^2}}, \tag{24.4}$$

or

$$c=\sqrt{\frac{e^{-\alpha t}(\operatorname{ch}\alpha t-\cos\beta t)}{2(\alpha^2+\beta^2)}}.$$

This equation then represents the desired expression for the dynamic characteristic of the resonator. The quantity c depends, as we can see, on the resonator parameters, i.e., on α and ω_0, as well as on the two independent variables ω and t. It is easy to affirm the fact that as $t\to\infty$, Eq. (24.4) tends in the limit to

$$c\xrightarrow[t\to\infty]{}\frac{1}{2}\frac{1}{\sqrt{\alpha^2+\beta^2}}=y. \tag{24.5}$$

This then is the expression for the static characteristic, i.e., the ordinary resonance curve.

We obtained the expression for the dynamic characteristic of the resonator from the time point of view, i.e., considering the steady-state regime of the resonator. The same problem could have been approached in purely spectral language as follows.

We shall solve the problem of the behavior of the resonator as a stationary one, i.e., we shall assume that the imposed potential is represented by a sum of sinusoidal potentials. Let us follow the procedure described in Section 13, i.e., let us apply to the resonator equation

$$L\frac{di}{dt}+Ri+\frac{1}{C}\int i\,dt=E(t)$$

a Fourier transform. We then obtain

$$\left(j\omega L + R + \frac{1}{j\omega C}\right) S_i = Z S_i = S_E,$$

where S_E and S_i are the potential and current spectra, respectively. Further,

$$i = \frac{1}{2\pi}\int_{-\infty}^{+\infty} S_i e^{j\omega t}\,d\omega = \frac{1}{2\pi}\int_{-\infty}^{+\infty} \frac{S_E}{Z} e^{j\omega t}\,d\omega.$$

However, in the case considered the spectral density is a function, not only of the frequency, but of the time. In fact, the imposed potential E (t) is assumed to be sinusoidal, but is switched into the analyzer at the instant t = 0. Consequently, for each given instant of time t the potential acting on the analyzer is represented by the following functions:

$$\begin{aligned} E(t) &= 0 && \text{for } t < 0, \\ E(t) &= \sin \Omega u && \text{for } 0 < u < t, \end{aligned}$$

and the spectrum of this function at the given instant, i.e., the running spectrum, is expressed by the equation

$$S_E = \int_0^t \sin \Omega u e^{-j\omega u}\,du = \\ = \frac{\Omega}{\Omega^2 - \omega^2}\left[1 - e^{-j\omega t}\left(\cos \Omega t + j\frac{\omega}{\Omega}\sin \Omega t\right)\right],$$

where ω is the running frequency of the spectrum, t is the passing time, u is the variable of integration. Thus we obtain for the current

$$i = \frac{\Omega}{2\pi}\int_{-\infty}^{+\infty} \frac{e^{j\omega t} - \left(\cos \Omega t + j\frac{\omega}{\Omega}\sin \Omega t\right)}{(\Omega^2 - \omega^2)\left[R + j\left(\omega L - \frac{1}{\omega C}\right)\right]}\,d\omega.$$

The integrand function has two simple conjugate poles below the real axis:

$$a = \pm \omega_1 + j\alpha.$$

(The points a = $\pm\Omega$ are not poles; of this we can be sure after resolving the indeterminateness obtained.) The residues of the integrand are equal to

$$\frac{\omega_1 \pm j\alpha}{2j\omega_1 L} \frac{e^{j(\pm\omega_1+j\alpha)t} - \left(\cos\Omega t + j\frac{\pm\omega_1+j\alpha}{\Omega}\sin\Omega t\right)}{\Omega^2-(\pm\omega_1+j\alpha)^2}.$$

Carrying out the computation, we find (having assumed L = 1)

$$i = \frac{\Omega}{(\Omega^2-\omega_0^2)^2+4\alpha^2\Omega^2}\{e^{-\alpha t}[(\Omega^2-\omega_0^2)\cos\omega_1 t -$$

$$-\frac{\alpha}{\omega_1}(\Omega^2+\omega_0^2)\sin\omega_1 t] - (\Omega^2-\omega_0^2)\cos\Omega t + 2\alpha\Omega\sin\Omega t\}.$$

This equation is the exact solution of the problem and agrees, of course, with the solution obtained in any other way. Computing the envelope, we find

$$c = \Omega\sqrt{\frac{1+\left(\frac{\omega_0}{\omega_1}\right)^2 e^{-2\alpha t} - 2\frac{\omega_0}{\omega_1}e^{-\alpha t}\cos(\Omega-\omega_1)t}{(\Omega^2-\omega_0^2)^2+4\alpha^2\Omega^2}}.$$

This expression differs somewhat from (24.4) as a result of the fact that Eq. (24.4) is not an exact one; it was obtained from the abridged Eq. (24.3). It is interesting to note that in the exact solution the cosine argument is simpler in form.

Let us proceed now to an investigation of the dynamic characteristic. In order for it to be possible to present a clear idea of the quantity c as a function of two variables ω and t, it is shown in relief in Fig. 37.

The curves lying in the planes parallel to $c0\frac{\omega}{\omega_0}$ are the dynamic resonance curves. The static resonance curve is also shown in the figure as the limit to which the dynamic curves tend.

The curves lying in the planes parallel to $c0\omega_0 t$ are curves showing the amplitude of the oscillations in the resonator for varying degrees of mistuning. These curves have an oscillating nature as a consequence of the beats between the driving frequency and the frequency of the natural oscillations of the resonator that are excited when it is turned on. The beat frequency is obviously greater when the mistuning is greater. When $\omega = \omega_0$, as expected, the process of settling follows an exponential law.

Examining Fig. 37, we see that the sharpness of the resonance curve depends on the time during which the resonator is turned on; the resonance sharpness and, consequently, the resolving power are greater when the operating time is longer. It is in just this way that we get the concept of the dynamic resolving power of a resonator. Unlike its static counterpart, the dynamic resolving power depends not only on the parameters of the resonator, but also on the time.

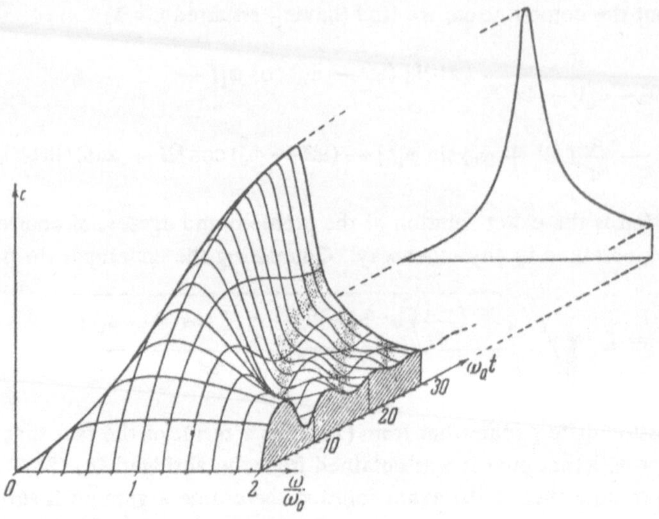

Fig. 37.

Let us derive a few relations. Let us assume that we have agreed to express the resonance sharpness by the width $\Delta\omega$ of the resonance curve at a height equal to $\frac{1}{\sqrt{2}}$ the maximum ordinate. Making use of (24.4), we note that the maximum amplitude is obtained when $\beta = 0$, so that

$$c_{\max} = \frac{1 - e^{-\alpha t}}{2\alpha}.$$

Composing the expression

$$\frac{c^2}{c_{\max}^2} = \frac{1}{2},$$

we obtain the equation

$$1 - \cos\beta t - \frac{1}{2}\left(\frac{\beta^2}{\alpha^2} - 1\right)(\operatorname{ch}\alpha t - 1) = 0.$$

As $t \to 0$ this equation assumes the form

$$1 - \cos \beta t - \frac{1}{4} \beta^2 t^2 = 0.$$

Solving this equation by the method of false position, for $\alpha = 0.1$ we obtain the solution cited below in the form of a table.

t	0	10	20	30
βt	2.78	2.87	3.13	3.57

Now we must go over from β to ω.

Assuming, as is the custom in electronic engineering, that the mistuning is small and the resonance curve symmetric,[1] we obtain

$$\beta = \frac{\omega_0^2 - \omega^2}{2\omega} = \frac{(\omega_0 + \omega)(\omega_0 - \omega)}{2\omega} \approx \omega_0 - \omega = \frac{\Delta\omega}{2},$$

i.e.,

$$2\beta \approx \Delta\omega.$$

Now we can assemble the following table:

t	0	10	20	30	∞
$\frac{\Delta\omega}{\omega_0}$	∞	0.575	0.313	0.238	0.200
$\Delta\omega \cdot t$	5.56	5.75	6.26	7.14	∞

[1]The indicated assumptions are not peremptory. The same result can be obtained in a more general way. We have the notation

$$\beta = \frac{|\omega_0^2 - \omega^2|}{2\omega},$$

(Footnote continued on next page)

The first line in Fig. 38a gives the relative width of the resonance curve and thus characterizes the dynamic resolving power of the resonator. With the passage of time $\frac{\Delta\omega}{\omega_0}$ tends to the constant limit d – this quantity expresses the relative width of the static resonance curve.

The second line (Fig. 38b) gives the product $\Delta\omega \cdot t$. When the values of t are small this is a constant quantity, and moreover – this is really quite remarkable – it does not depend on the attenuation of the resonator. With the passage of time the sharpness of the resonance curve tends to a constant limit, while the time increases without bound. Consequently, the curve for the dependence of $\Delta\omega \cdot t$ on t has an asymptotic straight line passing through the origin. The slope of this line is equal, of course, to d.

In order to clarify the concept of dynamic resolving power of the analyzer, in Fig. 39 a diagram is shown in relief, illustrating the evolution of the analyzer

([1]Continuation of footnote from previous page)

$$\frac{\beta}{\omega_0} = \frac{1}{2}\left(\frac{\omega_0}{\omega} - \frac{\omega}{\omega_0}\right).$$

Setting up and solving the resultant two quadratic equations with respect to $\epsilon = \frac{\omega}{\omega_0}$ we obtain the following, retaining only the meaningful roots:

$$\varepsilon_1 = \sqrt{\left(\frac{\beta}{\omega_0}\right)^2 + 1} - \frac{\beta}{\omega_0},$$

$$\varepsilon_2 = \sqrt{\left(\frac{\beta}{\omega_0}\right)^2 + 1} + \frac{\beta}{\omega_0}.$$

Therefore,

$$\varepsilon_2 - \varepsilon_1 = \frac{\omega_2 - \omega_1}{\omega_0} = \frac{\Delta\omega}{\omega_0} = \frac{2\beta}{\omega_0},$$

from which, as before,

$$\Delta\omega = \beta.$$

In this derivation we have assumed neither smallness of the mistuning nor symmetry of the resonance curve.

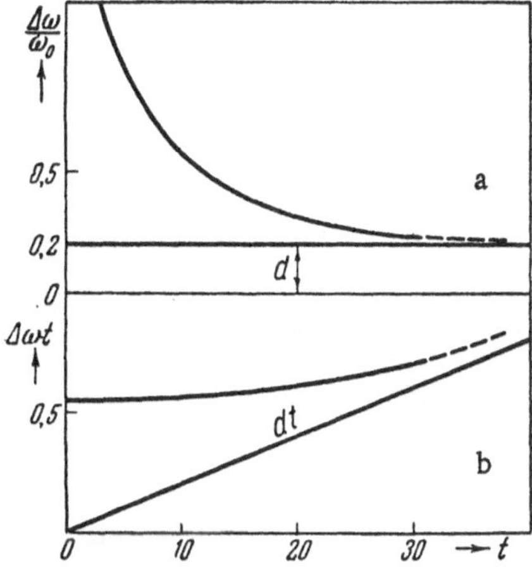

Fig. 38.

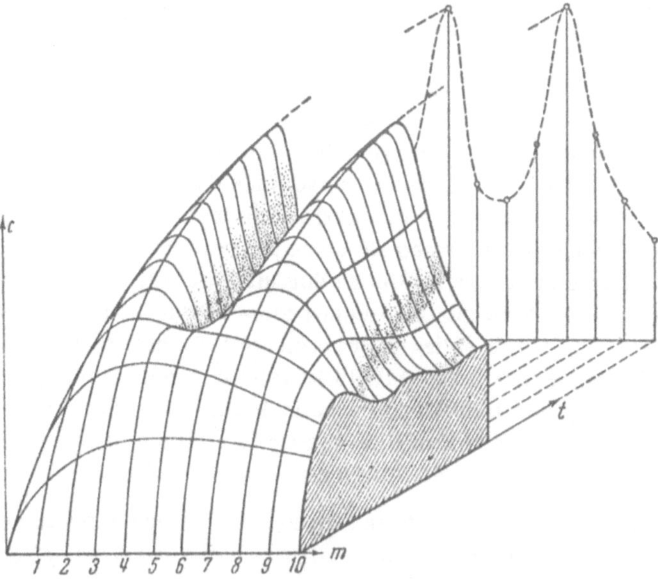

Fig. 39.

reading with the passage of time.[1] It is assumed that the analyzer is an assembly of resonators with identical attenuation and with equally spaced resonance frequencies. At the time t = 0 two sinusoidal oscillations are delivered to the analyzer. As evident from the figure, in the initial stage of the effect the analyzer does not separate these oscillations. Only after the passage of a certain amount of time does the saddle begin to form, and then gradually deepen. In the limit the steady-state analyzer reading, whose envelope has the form of a double-humped curve (see Section 19), is obtained.

25. Dynamic Characteristic of a Resonator Under the Action of a Variable Frequency

In the preceding section the problem of the dynamic resolving power of a resonator in a system of simultaneous analysis was considered.

Now we wish to develop relations for consecutive analysis, i.e., in the case when the resonator excitation frequency is varied.

Under the action of a time-varying frequency on the resonator, resonance effects are observed. However, if the variation in frequency proceeds with sufficient rapidity, then the most intense oscillations will not occur at the instant when the excitation frequency coincides with the resonance frequency, but somewhat later, since the resonator is not able to oscillate in immediate sympathy; for this very reason the greatest oscillation amplitude turns out to be less than for resonance in the steady-state regime. If we construct a graph showing the dependence of the oscillation amplitude on the instantaneous frequency, the dynamic resonance curve will be obtained. This curve can differ substantially from the static resonance curve when the frequency is varied rapidly, to wit: 1) the position of the maximum is shifted along the frequency scale (corresponding to a displacement along the time scale in the direction of delay); 2) the height of the maximum diminishes; 3) the curve becomes asymmetric—the left-hand incline becomes more sloping; 4) the transmission band increases (i.e., the resolving power decreases); 5) the effect becomes complicated by more and more rapid fluctuations in amplitude due to beats between the driving oscillations and the natural oscillations that arise upon passing through resonance. All of these features are visible in Fig. 40, in which the static (a) and dynamic (b) resonance curves are shown.

The theory of the effect has been the subject of numerous investigations (4, 7, 19, 23, 26]. The following presentation is based on papers by Turbovich

[1]Figure 39 is constructed from the same data as Fig. 37. However, the details are omitted and the entire relief is smoothed over somewhat, so that the figure reproduces the effect only in its general features.

[15, 16], which give a rather general approach to the problem, at the same time making it possible to obtain answers to the problems that stem from the demands of practical application.

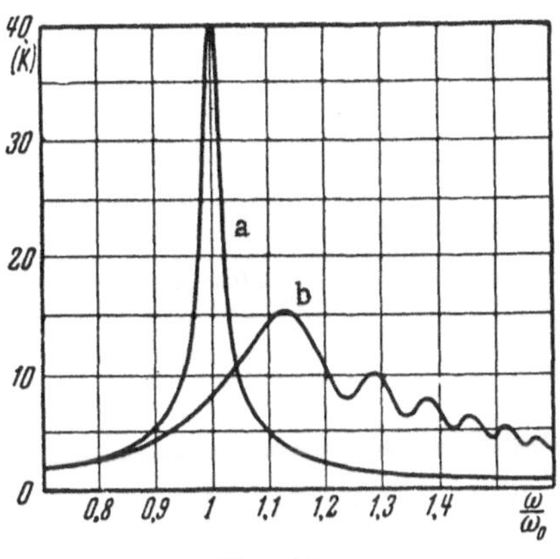

Fig. 40.

Let the frequency be varied with time according to the linear law

$$\omega = \omega_1 + \lambda t. \qquad (t > 0)$$

The potential acting on the resonator can be written in complex form:

$$U_1 = e^{j\vartheta(t)},$$

where

$$\vartheta(t) = \int_0^t \omega\, dt = \omega_1 t + \frac{1}{2}\lambda t^2.$$

The output potential is described by the Duhamel integral

$$U_2 = \int_0^t g(\tau)\, U_1(t-\tau)\, d\tau = \int_0^t g(\tau)\, e^{j\vartheta(t-\tau)}\, d\tau,$$

where $g(\tau)$ is the pulse reaction of the resonator. Substituting the value for ϑ, we find

$$U_2 = e^{j\left(\omega_1 t + \frac{1}{2}\lambda t^2\right)} \int_0^t g(\tau)\, e^{-j\left(\omega_1\tau + \lambda t\tau - \frac{1}{2}\lambda\tau^2\right)}\, d\tau.$$

But

$$e^{j\left(\omega_1 t + \frac{1}{2}\lambda t^2\right)} = U_1; \qquad \omega_1 + \lambda t = \omega,$$

so that

$$U_2 = U_1 \int_0^t g(\tau)\, e^{-j\left(\omega\tau - \frac{1}{2}\lambda\tau^2\right)}\, d\tau.$$

The dynamic transmission coefficient is defined as the ratio of the complex potentials U_2 and U_1:

$$K_д = \frac{U_2}{U_1} = \int_0^t g(\tau)\, e^{-j\omega\tau}\, e^{j\frac{1}{2}\lambda\tau^2}\, d\tau. \qquad (25.1)$$

This is an exact equation having a rather universal character, since it is suitable for any resonator.

Let us turn now to approximations. First of all, we shall note that $g(t)$ is a function containing an exponentially decreasing factor. Therefore, introducing only an insignificant error (which is evaluated below), we can write the upper limit at infinity:

$$K_d \simeq \int_0^\infty g(\tau)\, e^{-j\omega\tau}\, e^{j\frac{1}{2}\lambda\tau^2}\, d\tau. \qquad (25.2)$$

When $\lambda \to 0$, i.e., upon passing to the static regime, we obtain the familiar expression for the static transmission coefficient as the Fourier transform of the pulse reaction:

$$K = \int_0^\infty g(\tau)\, e^{j\omega\tau}\, d\tau. \qquad (25.3)$$

Thus the difference between the dynamic and static transmission coefficients is equal to

$$\Delta K = K_d - K = \int_0^\infty g(\tau)e^{-j\omega\tau}\left(e^{j\frac{1}{2}\lambda\tau^2} - 1\right)d\tau. \qquad (25.4)$$

We note further that there is an interesting case when the dynamic resonance curve differs but slightly from the static version, since only then does the analyzer error, computed from the static parameters, become small. Assuming the argument $\lambda\tau^2$ to be small for all values of τ, for which the diminishing function $g(\tau)$ still has a significant magnitude, we can expand the expression in the parentheses after the integral sign in (25.4) in a power series and limit it to the first two terms of the expansion as a first approximation. This will give

$$\Delta K \simeq \int_0^\infty g(\tau)e^{-j\omega\tau}\left(j\frac{1}{2}\lambda\tau^2 - \frac{1}{8}\lambda^2\tau^4\right)d\tau.$$

We now note that the appearance of the factors τ^n can be thought of as the result of $\underline{n}$-fold differentiations with respect to the parameter ω.

Consequently,

$$\Delta K \simeq -j\frac{1}{2}\lambda\frac{d^2}{d\omega^2}\int_0^\infty g(\tau)e^{-j\omega\tau}d\tau - \frac{1}{8}\lambda^2\frac{d^4}{d\omega^4}\int_0^\infty g(\tau)e^{-j\omega\tau}d\tau,$$

in other words,

$$\Delta K \simeq -j\frac{1}{2}\lambda K'' - \frac{1}{8}\lambda^2 K^{IV}.$$

Thus, for the dynamic transmission coefficient we get

$$K_d = K + \Delta K = K - j\frac{1}{2}\lambda K'' - \frac{1}{8}\lambda^2 K^{IV}. \qquad (25.5)$$

Assuming

$$K = A + jB, \quad C = |K|^2 = A^2 + B^2,$$

we can write a general expression for the square of the modulus of the dynamic transmission coefficient:

$$C_{\text{d}} = |K_{\text{d}}|^2 = C + \lambda(AB'' - BA'') + \\ + \frac{1}{4}\lambda^2 [(A'')^2 + (B'')^2 - AA^{IV} - BB^{IV}]. \tag{25.6}$$

The approximate Eqs. (25.5) and (25.6) are of a general nature. In what follows we shall consider only a simple resonator, after the circuit of Fig. 41, for which

$$K = \frac{1}{1 - \frac{\omega^2}{\omega_0^2} + j\frac{\omega}{\omega_0}d} \simeq \frac{1}{-\varepsilon + jd} = \frac{1}{jd}\frac{1}{1 + jx},$$

where

$$x = \frac{\varepsilon}{d} = \frac{2(\omega - \omega_0)}{\omega_0 d}$$

is the generalized mistuning. Correct to a constant multiplier

$$K = \frac{1}{1 + jx} \tag{25.7}$$

and we shall avail ourselves of this expression in later discussions.

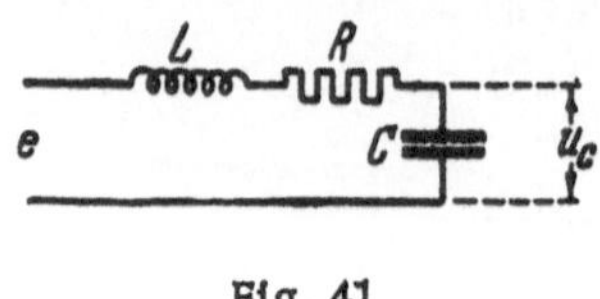

Fig. 41.

Let us subject the problem to further investigation. Bearing in mind the practical issues of the calculation and construction of the analyzers (as well as curve-tracers and other instruments using a frequency-modulated testing potential), we obviously should explain in what way any of the parameters defining the resonance curve depend on the rate of change of frequency. In particular, we are interested in the following: 1) the displacement of the maximum $\underline{s}$; 2) the drop in the height of the maximum ΔC; 3) the displacement of the transmission band $\frac{1}{2}(s_1 + s_2)$; 4) the change in width of the transmission band $s_2 - s_1$. The meaning of the notation introduced here is clarified in Fig. 42, in which, as before, $\underline{a}$ is the static, $\underline{b}$ the dynamic resonance curve.

In order to determine the numerical values we shall proceed from (25.5), which expresses the dynamic transmission coefficient, rewriting it in terms of new variables:

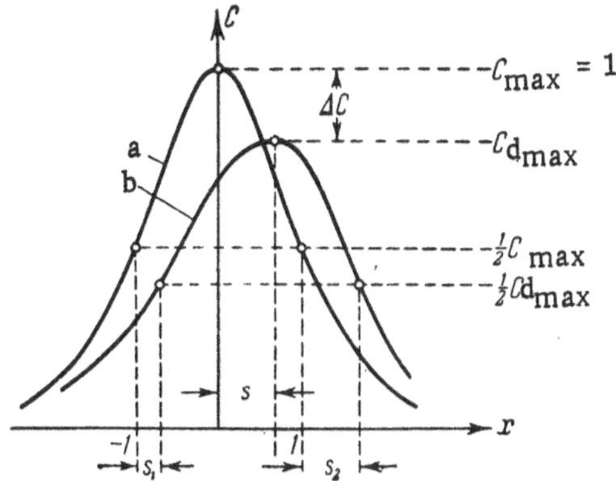

Fig. 42.

$$K_d = K - j\frac{1}{2}\lambda\frac{d^2K}{d\omega^2} - \frac{1}{8}\lambda^2\frac{d^4K}{d\omega^4} =$$

$$= K - j\frac{1}{2}\lambda\left(\frac{dx}{d\omega}\right)^2\frac{d^2K}{dx^2} - \frac{1}{8}\lambda^2\left(\frac{dx}{d\omega}\right)^4\frac{d^4K}{dx^4}$$

or

$$K_d = K - j\mu K''_x - \frac{1}{8}\mu^2 K^{IV}_x, \tag{25.8}$$

where

$$\mu = \lambda\left(\frac{dx}{d\omega}\right)^2 = \frac{4\lambda}{\omega_0^2 d^2}.$$

In Eq. (25.8) K = K (x) and the derivatives are taken with respect to the argument $\underline{x}$. For the modulus of the dynamic transmission coefficient we find

$$|K^2_{d}| = C_d = C - 4\mu x C^3 - \mu^2(5C^3 - 48x^2C^5). \tag{25.9}$$

The maximum for the static transmission coefficient is attained when x = 0:

$$C_{max} = C(0) = 1; \qquad C'(0) = 0.$$

The maximum for the dynamic transmission coefficient occurs for some value of x = s, which is what we wish to find. Let us expand C'_d in a Maclaurin series, limiting it to the first-order terms:

$$C'_d(s) = C'_d(0) + sC''_d(0) = 0.$$

Substituting here the value of C_d from (25.9) and retaining only terms as high as μ, we obtain an equation in s; the solution of this equation is

$$s = 2\mu, \tag{25.10}$$

i.e., the displacement of the maximum with dynamic resonance is of the same order of smallness as the parameter μ, which is proportional to the rate of change of frequency.

In order to find the relative change in the height of the maximum we write the expression

$$\Delta C = C_d(s) - C(0) = C_d(0) + sC'_d(0) - C(0).$$

Substituting here the values of C_d and s from (25.9) and (25.10), we find, correct to μ^2,

$$\Delta C = -\mu^2. \tag{25.11}$$

Therefore, the change in height of the maximum is a second-order small quantity relative to μ.

Let us now determine the values of the generalized mistuning x for which the square of the modulus of the transmission coefficient is equal to half the maximum value (in correspondence with the conventional definition for transmission bandwidth). For the static transmission coefficient

$$C(\pm 1) = \frac{1}{2} C(0) = \frac{1}{2}.$$

For the dynamic transmission coefficient

$$C_d(-1 + s_1) = C_d(1 + s_2) = \frac{1}{2} C_d(s) = \frac{1}{2}(1 - \mu^2).$$

Expanding this expression in a series with respect to the desired small parameters s_1 and s_2 (correct to μ^2), we obtain an equation for determining these quantities. The difference $s_2 - s_1$ gives the increment in the relative transmission bandwidth; the computation gives

$$s_2 - s_1 = \frac{5}{2}\mu^2. \tag{25.12}$$

The broadening of the transmission band, as we can see, is of second-order smallness relative to μ. The center of the transmission band for the dynamic resonance curve is shifted relative to x = 0 by an amount equal to half the sum of s_1 and s_2. The computation gives

$$\frac{1}{2}(s_1 + s_2) = \mu. \tag{25.13}$$

Comparing (25.13) and (25.10), we note that the displacement of the center of the band is half as large as the displacement of the maximum. This is due to the aforementioned asymmetry in the dynamic resonance curve.

Thus the single-term expressions of the first approximation for the pertinent parameters of the dynamic resonance curve are proportional to μ and μ^2, i.e., they are first- or second-order small quantities when μ is a first-order small quantity. The gross distortion of the resonance curve as shown in Fig. 42 would hardly be encountered in practice in constructing electronic measuring apparatus; the dynamic resonance curve in Fig. 42 refers to a value of $\mu \simeq 30$ (d = 0.025).

Let us turn now from the generalized to the original variables. We shall designate the transmission bandwidth by $\Delta\omega$; the symbol δ will be used to designate the increments in any of the parameters in the dynamic regime. Then, instead of Eqs. (25.10), (25.11), and (25.12) we have

$$\frac{\delta\omega_0}{\omega_0} = \mu d = \frac{4\lambda}{\omega_0^2 d}, \tag{25.10'}$$

$$\frac{\delta C_{\max}}{C_{\max}} = -\mu^2 = -\frac{16\lambda^2}{\omega_0^4 d^4}, \tag{25.11'}$$

$$\frac{\delta(\Delta\omega)}{\Delta\omega} = \frac{5}{4}\mu^2 = \frac{20\lambda^2}{\omega_0^4 d^4}. \tag{25.12'}$$

The necessary analysis time (or period of frequency modulation when the form of the modulation is a sawtooth) can be determined from the equation

$$T = \frac{2}{\pi}\frac{F}{\mu(\Delta f)^2},$$

where F is the modulation sweep, i.e., the full width of the frequency range, Δf is the resolving power, i.e., the width of the transmission band (in the derivation of this equation it was assumed that $\frac{\Delta f}{f_0} = d$).

Thus, e.g., for a heterodyne analyzer operating at acoustic frequencies ($F = 10$ kc) with an electromechanical resonator tuned to a frequency of 50 kc, given a resolving power of 25 cps (corresponding to $d = 5 \cdot 10^{-4}$), assuming $\mu = 0.1$ (i.e., a permissible broadening of the band by an amount of the order 1%), we obtain

$$T = \frac{2}{\pi} \frac{10^4}{0.1\,(25)^2} \simeq 100 \text{ sec.}$$

In practice it is considered sufficient to set forth requirements that are less rigid: it is assumed that $\mu \leq 1$. When μ is of the order of unity the approximate equations given above are, of course, no longer applicable.

Let us turn now to the original expression (25.1) for the dynamic transmission coefficient. In this expression we replaced the upper limit by infinity [see (25.2)], assuming that the error introduced by this substitution was small. We must now estimate the magnitude of this error, which is equal to

$$R = \int\limits_{t}^{\infty} g(\tau)\, e^{-j\omega\tau}\, e^{j\frac{1}{2}\lambda\tau^2}\, d\tau.$$

The modulus of the integral is less than the integral of the modulus; therefore,

$$|R| < \int\limits_{t}^{\infty} |g(\tau)|\, d\tau.$$

Substituting

$$g(t) = \omega_0 e^{-\alpha t} \sin \omega_0 t,$$

we write

$$|R| < \omega_0 \int\limits_{t}^{\infty} e^{-\alpha\tau}\, d\tau = \frac{2}{d}\, e^{-\alpha t}.$$

But

$$\alpha t = \alpha \frac{\omega - \omega_1}{\lambda} = \frac{x - x_1}{\mu}$$

and, accordingly,

$$|R| < \frac{2}{d} e^{-\frac{x - x_1}{\mu}}.$$

The relative error is obtained by referring R to $K_{max} = \frac{1}{jd}$

$$\left|\frac{R}{K_{max}}\right| = 2e^{-\frac{x - x_1}{\mu}}.$$

All the calculations have been made with an accuracy to μ^2. In order that the relative error be neglected, it is sufficient that it have the next order of smallness, i.e., that it be of the order μ^3. Rejecting the factor 2, which does not have any effect on the degree of smallness, we obtain

$$e^{-\frac{x - x_1}{\mu}} \leqslant \mu^3$$

or, after taking the logarithm of both sides,

$$x - x_1 \geqslant 3\mu \ ln \frac{1}{\mu}.$$

The meaning of this inequality is contained in the fact that in order to obtain sufficiently accurate results in the frequency band whose extent is determined by the generalized mistuning $\underline{x}$, it is necessary that the initial frequency corresponding to the value $\underline{x}$ be sufficiently removed from the boundary of the investigated band. Let, for example, $\mu = 0.1$. Then

$$x - x_1 \geqslant 3 \cdot 0.1 \cdot \ln 10 = 0.69.$$

We recall that the transmission bandwidth of a simple resonator, expressed in terms of the generalized mistuning, is equal to 2. Thus it is necessary to recede to the left of the edge of the investigated band only a fraction of the transmission band of the resonator, in order to be justified in neglecting the error R upon transition from (25.1) to (25.2). This condition, of course, in practice is always fulfilled, and all the relations derived on the basis of Eq. (25.2) are thus confirmed.

The foregoing outline has been based on the assumption that the difference between the dynamic frequency characteristic and the static frequency characteristic is small; it was considered self-evident that we should have a tendency toward such an assumption. This is valid when we consider the analyzer on the basis of the parameters of the static characteristic of the resonator. In this case the difference between the static and dynamic characteristics is considered a source of error, and it is thus reasonably desired that this error be small. But another approach to the problem is possible. Specifically, if the rate of change of the frequency is constant, then the analyzer can be treated on the basis of the parameters of the dynamic characteristic. In this case the rate of change of the frequency can be as large as we like, the only condition being that we agree upon the values of the dynamic parameters (for example, the transmission bandwidth).

In conclusion, it is interesting to note that the problem of the effects that arise under the action of a variable frequency on a resonance system is also a matter of concern for other branches of technology. Thus, in starting up machinery with nonuniform rotating masses on a flexible shaft, dangerous oscillations can arise upon passing through resonance (the so-called "critical speed"). The mathematical formulation of this problem has the same accuracy as the present one [7, 26]. However, the purpose of the calculation is the opposite: it is necessary to find safe starting conditions, i.e., to insure a rapid enough change in the speed of rotation, so to speak, so as to "jump through" resonance without exciting oscillations with an excessive amplitude at the critical speed. In our terminology this means that it is necessary, having specified the largest permissible value of $K_{d_{max}}$, to choose the least value of $\mu \gg 1$.

26. Analysis of Isolated Pulses

The analysis of nonperiodic effects, particularly pulses, both in setting up the problem and in visualizing the effects differs a great deal from the analysis of periodic effects.

In order to clarify these peculiarities we shall consider one possibility for the analysis of pulses. Let us recall that for the analysis of pulses the method of consecutive analysis is wholly unsuitable. Only the method of simultaneous analysis can be applied for the analysis of pulses (this already constitutes one of the differences between the analysis of pulses and the analysis of periodic effects).

Let us assume that we wish to apply an analyzer in the form of an assembly of resonators tuned to different frequencies for the purpose of analyzing pulses. Such an analyzer would be suitable for the analysis of periodic effects; its applicability for the analysis of pulses must still be established.

We shall first show that a lossless resonator assembly can given exact discrete values of the spectral density of the pulse. The spectrum of the pulse is essentially a continuous curve which represents the spectral density as a function of frequency. It is clear that an assembly consisting of a finite number of resonators can give a reading represented graphically only by a set of discrete points. The only hope is that these points will lie on the curve of a complex spectrum. The greater the number of points, i.e., the greater the number of resonators, the better will be the definition of the curve. This consideration gives some idea as to the way in which the problem of the resolving power of the analyzer is handled in the present case.

Let the equation of a resonator without attenuation be written in the form

$$U'' + \omega_k^2 U = \omega_k^2 a_k f(t), \tag{26.1}$$

where f(t) is the pulse to be analyzed, which we shall assume equal to zero outside the interval $0 < t < \tau$, k is the order number of the resonator, U is the output potential, a_k is a proportionality coefficient and is specified by the input divider.

The solution of (26.1) can be written in the form

$$U(t) = a_k \int_0^t g(t-u) f(u)\, du, \tag{26.2}$$

where g(t) is the time characteristic, i.e., the solution of (26.1) when the right-hand side has the form

$$f(t) = \delta(t).$$

For the case under consideration we have

$$g(t) = a_k \omega_k \sin \omega_k t,$$

where $\omega_k = \dfrac{1}{\sqrt{L_k C_k}}$ is the natural frequency of the kth resonator. Consequently,

$$U(t) = a_k \omega_k \int_0^t \sin \omega_k (t-u) f(u)\, du =$$

$$= a_k \omega_k \left[\sin \omega_k t \int_0^t f(u) \cos \omega_k u\, du - \cos \omega_k t \int_0^t f(u) \sin \omega_k u\, du \right].$$

Let us determine the envelope of this oscillation; it will be expressed in terms of the sum of the squares of the coefficients associated with $\sin \omega_k t$ and $\cos \omega_k t$:

$$c = a_k\omega_k \left[\left(\int_0^t f(u) \cos \omega_k u \, du \right)^2 + \left(\int_0^t f(u) \sin \omega_k u \, du \right)^2 \right]^{\frac{1}{2}}. \quad (26.3)$$

This quantity (which is the variable "amplitude" of the oscillation) is a function of time; the time dependence is determined by the fact that t appears in the upper limits of the integrals.

But after $t = \tau$ the pulse f(t), and with it the integrand, revert to zero, and, consequently, afterward, i.e., for $t > \tau$, the quantity ξ retains the constant value that it took on at the instant $t = \tau$; the corresponding value is designated by c_τ. Physically, this means that after the external action on the resonator is finished, the amplitude of its oscillation thereafter remains unchanged (we recall that we are still considering a lossless resonator). But in such a case we can write the expression for c_τ with any constant limits of integration that contain the interval $0 - \tau$; in particular, even with infinite limits:

$$c_\tau = a_k\omega_k \left[\left(\int_{-\infty}^{+\infty} f(u) \cos \omega_k u \, du \right)^2 + \left(\int_{-\infty}^{+\infty} f(u) \sin \omega_k u \, du \right)^2 \right]^{\frac{1}{2}} . \quad (26.4)$$

In the integrals on the right-hand side it is easy to recognize the expressions for the sine and cosine components of the spectrum, so that

$$c_\tau = a_k\omega_k \sqrt{a^2 + b^2} = a_k\omega_k \Phi(\omega). \quad (26.5)$$

If now we select the factor

$$a_k\omega_k$$

to be identical for all the resonators, then it is found that the resonator oscillatory amplitudes observed when the action of the pulse is finished (correct to a constant multiplier) give exact discrete values of the spectrum for the frequencies $\omega = \omega_k$.

As for the interval $0 < t < \tau$, during this period the amplitudes of the resonator oscillations give the running spectrum of the pulse, as evident from Eq. (26.3).

It must be noted that the considered analyzer, consisting of <u>lossless</u> resonators and giving, as we have just been convinced, an ideal analysis of the pulse, in principle is unsuitable for the analysis of periodic effects. Essentially, for a resonator with no attenuation the steady-state regime does not in general exist, but the analysis of periodic effects is conceivable only in the steady state regime. This then is a manifestiation of still another substantial difference between the analysis of periodic and pulse effects.

It should be further noted that the analysis time for a pulse is determined by the duration of the pulse and in no way depends on the properties of the analyzer. Consequently, the relation considered in Section 23 between the analysis time and the resolving power of the analyzer is inapplicable to the analysis of pulses. Here again we have a difference between the analysis of periodic and pulse effects.

Let us now consider the way in which the relations must be modified if we are to assemble a pulse analyzer from real resonators with attenuation. The equation of the resonator in this case is the following:

$$\ddot{U} + 2\alpha\dot{U} + \omega_k^2 \dot{U} = a_k\omega_k^2 f(t). \tag{26.6}$$

The time characteristic has the form

$$g(t) = \omega_k e^{-\alpha_k t} \sin \omega_k t.$$

The solution of Eq. (26.6) is

$$U(t) = a_k\omega_k \int_0 e^{-\alpha_k (t-u)} \sin \omega_k (t-u) f(u)\, du =$$

$$= a_k\omega_k e^{-\alpha_k t} \left[\sin \omega_k t \int_0^t e^{\alpha_k u} f(u) \cos \omega_k u\, du - \right.$$

$$\left. - \cos \omega_k t \int_0^t e^{\alpha_k u} f(u) \sin \omega_k u\, du \right].$$

When t > τ we can write for the envelope

$$c = a_k\omega_k e^{-\alpha_k t} \left[\left(\int_{-\infty}^{+\infty} e^{\alpha_k u} f(u) \cos \omega_k u\, du \right)^2 + \right.$$
$$\left. + \left(\int_{-\infty}^{+\infty} e^{\alpha_k u} f(u) \sin \omega_k u\, du \right)^2 \right]^{\frac{1}{2}} \tag{26.7}$$

Comparing (26.7) and (26.4), we note the following. First of all, in the present instance the amplitude for $t > \tau$ does not remain constant in value, but falls off due to the factor $e^{-\alpha_k t}$, since the free oscillation is now damped. This means that the analyzer reading depends on the moment at which it is taken (naturally, after $t = \tau$). However, if we make the damping factor α_k the same for all the resonators, then with the passage of time the spectrum will not be distorted; only the scale factor will change, which is unimportant as long as the readings on each resonator are made simultaneously. Secondly, the presence of attenuation is manifested in the fact that in the integrands the weighting factor $\alpha^{\alpha u}$ appeared. Its least value is equal to $e^{\alpha\tau}$, this value differs less from unity when the product $\alpha\tau$ is smaller, i.e., when the attenuation is less, and when the pulse is shorter.

In a number of cases it may prove possible to assume approximately that

$$e^{\alpha u} \approx 1,$$

i.e., to neglect the influence of attenuation and thus reduce the case considered to the former one. If, on the other hand, this is not possible, then the analyzer reading will not correspond to the true spectrum of the pulse, and the problem consists in estimating the propagated error.

From Eq. (26.7) it is evident that the analyzer does not give us the spectrum of the function f(t), but of the modified (weighted) function

$$f_\alpha(t) = e^{\alpha t} f(t).$$

But if the spectrum of the function f(t) is

$$S(\omega) = \int_{-\infty}^{+\infty} f(t)\, e^{-j\omega t}\, dt,$$

then the spectrum of the modified function $f_\alpha(t)$ will be

$$S_\alpha = \int_{-\infty}^{+\infty} e^{\alpha t} f(t)\, e^{-j\omega t}\, dt = S(\omega + j\alpha).$$

Therefore, instead of the true values $\Phi(\omega_k)$, from the analyzer we read the values

$$\Phi_\alpha = |S(\omega_k + j\alpha)|. \tag{26.8}$$

From this equation the spectral distortion caused by the presence of attenuation can be calculated for a given specific pulse.

As an example let us take the analysis of a rectangular pulse with a duration τ. The analytic expression of this pulse can be written in the form

$$f(t) = \sigma(t) - \sigma(t - \tau).$$

Thus

$$\frac{1}{a_k} U(t) = \int_0^t g(t-u) f(u)\, du = \int_0^t g(t-u)[\sigma(u) - \sigma(u-\tau)]\, du =$$

$$= \begin{cases} \int_0^t g(t-u)\, du = \int_0^t g(u)\, du & 0 < t < \tau, \\ \int_0^\tau g(t-u)\, du = \int_{t-\tau}^t g(u)\, du & \tau < t < \infty. \end{cases}$$

Substituting here the value

$$g(t) = \omega_k e^{-\alpha_k t} \sin \omega_k t,$$

carrying out the integration, and assuming $\alpha_k \ll \omega_k$, we obtain the approximate solution

$$\frac{1}{a_k} U(t) = \begin{cases} 1 - e^{-\alpha_k t} \cos \omega_k t & 0 < t < \tau, \\ e^{-\alpha_k t} [e^{\alpha_k \tau} \cos \omega_k (t-\tau) - \cos \omega_k t] & \tau < t < \infty. \end{cases} \tag{26.9}$$

Assuming $a_k \omega_k = 1$, $\alpha_k = \alpha$, after some not too complicated computations we obtain the following for the envelope:

$$c(t) = \begin{cases} \frac{1}{\omega_k}(1 + e^{-\alpha t}) & 0 < t < \tau, \\ \frac{1}{\omega_k} M e^{-\alpha t} & \tau < t < \infty, \end{cases} \quad (26.10)$$

where

$$M = \sqrt{1 - 2e^{\alpha\tau}\cos\omega_k\tau + e^{2\alpha\tau}}. \quad (26.11)$$

The physical picture of the effect is this: when $t = 0$ $f(t)$ changes discontinuously from 0 to 1, as a result of which free oscillation of the resonator takes place; when $t = \tau$ $f(t)$ changes discontinuously in the opposite direction, i.e., from 1 to 0. In this case exactly the same oscillation arises, but with the opposite sign and with a time delay of τ (Fig. 43a).

The amplitude of the resulting oscillation that sets in after $t = \tau$ (Fig. 43b) depends understandably on the relative phase of both free oscillations; the phase shift depends in turn both on the frequency ω_k and on the interval τ. This is why the quantity M is a function of the argument $\omega_k\tau$. This dependence then determines the ability of the analyzer to discriminate in the spectral relations.

We note that the oscillation $\underline{x}$ is continuous, but the envelope $\underline{c}$ suffers a discontinuity at $t = \tau$. The upper equation in (26.10) includes a constant component; the envelope according to Eq. (26.10) is marked in Fig. 43b.

For the spectrum of a rectangular pulse of duration τ existing between $-\frac{\tau}{2}$ and $+\frac{\tau}{2}$, we had (see Section 11).

$$S(\omega) = \frac{\sin\omega\frac{\tau}{2}}{\omega\frac{\tau}{2}}.$$

In our problem the pulse is specified over the interval $0 < t < \tau$; consequently, for its spectrum we have

$$S(\omega) = e^{-j\omega\frac{\tau}{2}}\frac{\sin\omega\frac{\tau}{2}}{\omega\frac{\tau}{2}}.$$

Let us apply Eq. (26.8) for obtaining the distorted spectrum

$$\Phi_\alpha = |S(\omega + j\alpha)| = \left| e^{\frac{\tau}{2}(-j\omega+\alpha)} \frac{\sin\frac{\tau}{2}(\omega + j\alpha)}{\frac{\tau}{2}(\omega + j\alpha)} \right|,$$

which, after performing the computation, gives

$$\Phi_\alpha = \frac{1}{\omega\tau}\sqrt{1 - 2e^{\alpha\tau}\cos\omega\tau + e^{2\alpha\tau}}.$$

We see that, as it should be, the quantity M [see Eqs. (26.10) and (26.11)] expressed the observed spectrum directly. As for the spectral distortion, it is manifested most noticeably in the fact that, instead of having the zero that the true spectrum has at $\omega\tau = 2n\pi$, the distorted spectrum Φ_α, for the same arguments, has the finite values

$$\Phi_{\alpha_{\min}} = \frac{1 - e^{\alpha\tau}}{2n\pi}.$$

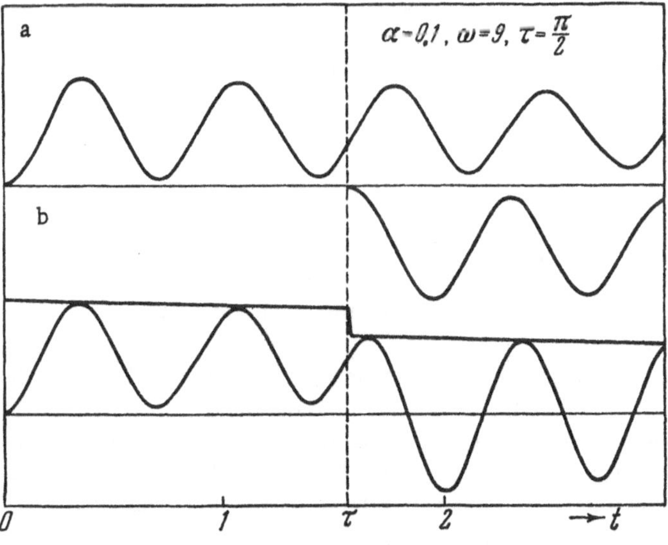

Fig. 43.

It is very interesting to note that before t = τ the analyzer still, so to speak, "does not know" with what it is dealing. At the beginning of the pulse it receives an impulse, which stimulates oscillations in all the resonators. Afterward any additional disturbance on the resonators will not bring about a reaction until $t = \tau$. Consequently, the analyzer detects the rectangular pulse as a unit function, and its reading gives the spectrum of this function, as evident from the upper equation in (26.10). When $t = \tau$ the position is suddenly determined; the pulse is formed and terminated, and from this instant the analyzer gives the spectrum of the pulse.[1]

Let us take a numerical example. Let the analyzer consist of 10 resonators with the following data: $\omega_k = k\omega_0$, $\omega_0 = 1$, $\alpha_k = \alpha = 0.1$, let the duration of the rectangular pulse be $\tau = \frac{\pi}{2}$. Then the zero in the spectrum will appear at the values $k = 4$ and $k = 8$. The factor M has the following values:

$$M = \begin{cases} \sqrt{1 + e^{2\alpha\tau}} = 1.54 & \text{for } k = 1,\ 3,\ 5,\ 7,\ 9, \\ 1 + e^{\alpha\tau} = 2.17 & \text{for } k = 2,\ 6,\ 10, \\ |1 - e^{\alpha\tau}| = 0.17 & \text{for } k = 4,\ 8. \end{cases}$$

A relief of $c(k,t)$ is constructed in Fig. 44 on the basis of Eq. (26.10). In this figure the amplitude is shown as a function of the resonator order number and time. When $t < \tau$ the analyzer gives a hyperbolic spectrum of the unit function. When $t = \tau$ the amplitude changes discontinuously; at this instant the spectrum of the rectangular pulse is formed. As we can see, the amplitudes of the resonators $k = 4$ and $k = 8$ are not equal to zero (distorted spectrum). The true spectrum of a rectangular pulse is shown in the appropriate scale in the foremost plane. In general the analyzer reproduces the spectrum quite satisfactorily.

Since the indentical value of the attenuation α was chosen for all the resonators, any cross section of the relief in a plane parallel to $c0k$ (i.e., a picture of the analyzer reading at any instant) gives a figure that differs only in its vertical scale.

[1]The general picture of the effect is clear, but it still remains to be shown why when $t < \tau$ the analyzer does not give the running spectrum of a rectangular pulse, as would be expected on the basis of Eq. (26.3). The essence of the matter is that in calculating the "envelope" from this equation we actually would obtain the running spectrum of the rectangular pulse, where it would turn out that the "amplitude" $\underline{c}$ varies with a frequency $\frac{1}{2}\omega_k$, i.e., with a frequency of the same order as the oscillation frequency of the resonator. It is impossible to observe the amplitude variation under such conditions and, consequently, such an investigation of the problem in the case of the present example would be formalistic.

Thus we have been assured that an analyzer in the form of an assembly of resonators can be successfully applied for the analysis of pulses.[1]

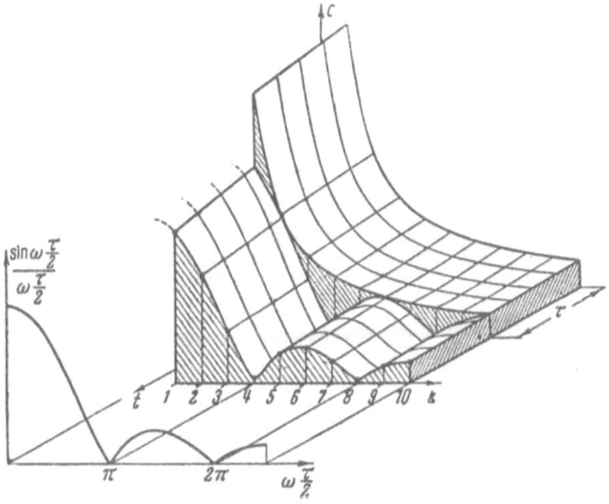

Fig. 44.

After a detailed examination of the pattern of the effects taking place in this case, we were able to notice a number of specific features of the analysis of pulses.

It is possible to approach the analysis of pulses by means of resonators in another way.

We note first of all that the resonator amplitude, which becomes steady at the instant of termination of the pulse, is a measure of the oscillatory energy stored by the resonator; the energy is proportional to the square of the amplitude. From this the possibility of modifying the means of taking the reading from the analyzer follows, namely: the resonator can be shorted by a resistive element, in which the energy stored in the resonator is dissipated, and by some means this energy can be measured.

[1]Rimskii-Korsakov also arrived at this result [12]. It must be noted that in his paper he follows an interesting and, so to speak, synthetic approach to the problem, first imposing certain demands for an ideal analyzer, then finding its physical characteristic. For the application considered this ideal analyzer turns out to be nothing other than the lossless resonator.

However, it is possible to adopt another approach, with the same beginning, to measure the loss of energy in the resonator. It turns out that upon termination of the process this energy is proportional to a known degree of approximation to the square of the spectral density of the pulse at the frequency of the resonator. We shall prove this.

Let us take right from the start a resonator with attenuation, for which the instantaneous value of the quantity observed (see the above) is

$$x(t) = a_k\omega_k \int_0^t e^{-\alpha_k(t-u)} \sin \omega_k (t-u)\, f(u)\, du =$$

$$= a_k\omega_k e^{-\alpha_k t}\left[\sin \omega_k t \int_0^t e^{\alpha_k u} f(u) \cos \omega_k u\, du - \right.$$

$$\left. - \cos \omega_k t \int_0^t e^{\alpha_k u} f(u) \sin \omega_k u\, du \right].$$

Let us write the expression for the energy expended by the time t:

$$E(t) = \int_0^t x^2\, dt.$$

The total energy given off in each resonator at the termination of both the pulse and the process within the resonator can be written in the form

$$E_k = a_k^2\omega_k^2 \int_0^\infty e^{-2\alpha_k t}\left[\sin \omega_k t \int_{-\infty}^{+\infty} f_\alpha(u) \cos \omega_k u\, du - \right.$$

$$\left. - \cos \omega_k t \int_{-\infty}^{+\infty} f_\alpha(u) \sin \omega_k u\, du \right]^2 dt.$$

We have here the notation

$$f_\alpha(u) = e^{\alpha_k u} f(u).$$

The internal integrals represent the spectral components A and B of the function f_α. Since these quantities do not depend on t, we have

$$E_k = a_k^2\omega_k^2 \left[a^2 \int_0^\infty e^{-2\alpha_k t} \sin^2 \omega_k t\, dt + b^2 \int_0^\infty e^{-2\alpha_k t} \cos^2 \omega_k t\, dt - \right.$$

$$\left. - ab \int_0^\infty e^{-2\alpha_k t} \sin 2\omega_k t\, dt \right] =$$

$$= \frac{a_k^2 \omega_k^2}{4\alpha_k} \left[a^2 + b^2 + \frac{\frac{d^2}{4}}{1+\frac{d^2}{4}} (b^2 - a^2) - \frac{d}{1+\frac{d^2}{4}} ab \right].$$

When the attenuation is small, i.e., when

$$\alpha_k \ll \omega_k, \quad d \ll 1,$$

we obtain approximately (neglecting the last two terms in the brackets)

$$E_k \approx \frac{a_k^2\omega_k^2}{4\alpha_k} (a^2 + b^2) = b_k \, \Phi_\alpha^2 \approx b_k \, \Phi^2,$$

since Φ_α for $d \ll 1$ differs little from Φ. If now we choose the parameters of the resonator in such a way that the quantity b_k does not depend on k, i.e., is identical for all resonators, then we find that the energy given off in each resonator is proportional (with an accuracy determined by the approximation made in the derivation) to the square of the spectral density corresponding to the frequency of a given resonator. This then is the required proof.

It is possible to devise various methods for taking the readings of the quantity E_k. It would be possible, for example, to connect a square-law electrodynamic ballistic instrument to each resonator. The integration could be done, not in the mechanical part (i.e., making use of the ballistic reading), but by means of an electrical integrating circuit. The integration could be done by employing the thermal inertia of a thermal device. Finally, the indicator could be imagined in the form of the following apparatus: the resistive impedance of each resonator is made to form a heating coil, which is attached to the bulb of the thermometer (or with the bulb inside). The reading is made from the height of the columns in the thermometers, which, placed in a row, would give a visual picture of the spectrum. The time constant of such a thermometric indicator is very easily controlled by varying the specific heats and thermal insulation of the thermometers.

The described means of analysis essentially reduces to a measurement of the energy of a complex oscillation occurring over a frequency band passed by a given resonator. Obviously, the averaging can be made more grossly, applying band analysis of the energy, for which the resonators would have to be replaced by appropriate band filters.

CHAPTER III

THE SPECTRA OF RANDOM PROCESSES

27. Spectral Representation of Random Processes

Until now we have applied the spectral representation to one kind of definite time function or another. Now it becomes necessary for us to extend the spectral ideas to objects of an essentially different character—to random processes, which in engineering and mathematics play a large and ever increasing role.

By a random process we mean a function of a continuously changing argument t, the values of which represent quantities of chance.

From this definition it is immediately evident that if we compute the spectral density of a random process $\xi(t)$ by the usual equation

$$S(\omega) = \int_{-\infty}^{\infty} \xi(t) e^{-j\omega t} dt, \tag{27.1}$$

then the function $S(\omega)$ obtained will be a random function. We find that one of the possible realizations of a random process can be called the spectrum. However, under the conditions of actual observation, during the course of a certain random process we can obtain only the running spectrum of a given realization, i.e., the random function

$$S_T(\omega) = \int^{T} \xi(t) e^{-j\omega t} dt. \tag{27.2}$$

It would be desirable to introduce the definition of the spectrum of a random process in such a way that it might lead to a nonrandom function of the frequency.

A random process is defined by the probability distribution of the values of the quantity ξ for any instant t. For the subsequent theory not only the probability distribution functions themselves are important, but also the first two moments

$$M(\xi) = \int x\varphi(x)\,dx = \xi_0, \tag{27.3}$$

$$M[\xi(t_1)\,\xi(t_2)] = \iint x_1 x_2 \varphi(x_1, x_2)\,dx_1\,dx_2 = B(t_1, t_2), \tag{27.4}$$

where $\varphi(x)$ and $\varphi(x_1, x_2)$ are the one- and two-dimensional probability densities, respectively, which in general depend on the time. However, in this section we shall consider only stationary (in the broad sense) processes, for which both quantities in (27.3) and (27.4) do not depend on the time; the first quantity is simply a constant, called the mean value of the random process, the second quantity depends only on the difference $|t_2 - t_1| = \tau$ and is called the correlation function.

The stationary random processes that we shall consider here have an ergodic property, in that averages taken over a set with a probability of unity are equal to the time averages, taken for any realization. Denoting time averages by a superior line, we can write the condition for ergodicity in the form

$$\xi_0 = M(\xi) = \overline{\xi(t)} = \lim_{T\to\infty} \frac{1}{T} \int_0^T \xi(t)\,dt, \tag{27.5}$$

$$B(\tau) = M[\xi(t)\,\xi(t-\tau)] = \overline{\xi(t)\,\xi(t-\tau)} = \lim_{T\to\infty} \frac{1}{T} \int_0^T \xi(t)\,\xi(t-\tau)\,dt. \tag{27.6}$$

We note immediately that

$$B(0) = M[\xi^2(t)] = \overline{\xi^2(t)} = \lim_{T\to\infty} \frac{1}{T} \int_0^T \xi^2(t)\,dt. \tag{27.7}$$

Thus the dispersion of a random process is

$$D(\xi) = M(\xi^2) - M^2(\xi) = \overline{\xi^2(t)} - [\overline{\xi(t)}]^2 = B(0) - \xi_0^2.$$

In the physical sense ξ_0 is a constant component, $B(0)$ is the total power of the process, ξ_0^2 is the power of the constant component, the dispersion expresses the power of the variable component. But it happens most frequently that we have to deal with processes for which $\xi_0 = 0$, so that $B(0)$ expresses the power of the process directly.

Our further development is based a theorem of Khinchin, according to which the correlation function $B(\tau)$ can be written in the form

$$B(\tau) = \int_{-\infty}^{\infty} e^{j\omega\tau}\, dF(\omega), \tag{27.8}$$

where $F(\omega)$ is a nondecreasing bounded function. If the function $F(\omega)$ is differentiable, then, designating

$$\frac{dF}{d\omega} = \frac{1}{2} G(\omega),$$

in place of (27.8) we obtain

$$B(\tau) = \frac{1}{2} \int_{-\infty}^{\infty} G(\omega)\, e^{j\omega\tau}\, d\omega, \tag{27.9}$$

so that $G(\omega)$ is the ordinary Fourier transform for $B(\tau)$:

$$G(\omega) = \frac{1}{\pi} \int_{-\infty}^{\infty} B(\tau)\, e^{-j\omega\tau}\, d\tau. \tag{27.10}$$

The function $G(\omega)$ is then what we call the statistic spectrum of a random process, and Eq. (27.10) is the fundamental definition of this function. It should be noted that both $B(\tau)$ and $G(\omega)$ are even functions of their respective arguments. Therefore, (27.9) and (27.10) can be written in real form as two Fourier cosine-transforms

$$B(\tau) = \int_{0}^{\infty} G(\omega) \cos \omega\tau\, d\omega, \tag{27.11}$$

$$G(\omega) = \frac{2}{\pi} \int_{0}^{\infty} B(\tau) \cos \omega\tau\, d\tau. \tag{27.12}$$

The physical meaning of the function $G(\omega)$ is easily interpreted by letting $\tau = 0$ in (27.11). We obtain

$$B(0) = P = \int_0^\infty G(\omega)\, d\omega, \tag{27.13}$$

where P denotes the total power of the process. Equation (27.13) shows that the function $G(\omega)$ represents the spectral density of the power. On this basis $G(\omega)$ is often called the energy spectrum of the process $\xi(t)$. The power contained within the finite band between ω_1 and ω_2 can be determined by integration of $G(\omega)$ within the appropriate limits:

$$P_{12} = \int_{\omega_1}^{\omega_2} G(\omega)\, d\omega.$$

The statistic spectrum can be expressed in terms of the running spectrum of the realization:

$$S_T(\omega) = \int_0^T \xi(t)\, e^{-j\omega t}\, dt.$$

Let us write an expression for the energy of the process $\xi(t)$ dissipated during the time T:

$$E_T = \int_0^T \xi^2(t)\, dt = \frac{1}{\pi} \int_0^\infty |S_T(\omega)|^2\, d\omega.$$

This relation expresses the Rayleigh theorem for a finite interval of time T. The average power over the time T is obtained by dividing by T:

$$P_T = \frac{E_T}{T} = \frac{1}{\pi T} \int_0^\infty |S_T(\omega)|^2\, d\omega. \tag{27.14}$$

Generally speaking, this quantity depends on T, but for a stationary process it tends with increasing T to a constant limit, which then expresses the power of the process:

$$P = \lim_{T\to\infty} P_T = \frac{1}{\pi} \lim_{T\to\infty} \frac{1}{T} \int_0^\infty |S_T(\omega)|^2\, d\omega. \tag{27.15}$$

Comparing (27.13) and (27.15), we see that the statistic spectrum is connected with the running spectrum by the relation

$$G(\omega)=\frac{1}{\pi}\lim_{T\to\infty}\frac{|S_T(\omega)|^2}{T}. \tag{27.16}$$

Let us derive a few other equations for computing $G(\omega)$. Let us take Eq. (27.10) and substitute into it the value of the correlation function in correspondence with its definition [see (27.4)]:

$$B(\tau)=M[\xi(t)\xi(t-\tau)]=\int_{-\infty}^{\infty}\int_{-\infty}^{\infty}x_1x_2\varphi(x_1,x_2,\tau)\,dx_1\,dx_2.$$

In the expression so obtained

$$G(\omega)=\frac{1}{\pi}\int_{-\infty}^{\infty}e^{-j\omega\tau}\,d\tau\int_{-\infty}^{\infty}\int_{-\infty}^{\infty}x_1x_2\varphi(x_1,x_2,\tau)\,dx_1\,dx_2$$

we first carry out the integration with respect to τ. Denoting

$$\psi(x_1,x_2,\omega)=\int_{-\infty}^{\infty}\varphi(x_1,x_2,\tau)\,e^{-j\omega\tau}\,d\tau,$$

we can write

$$G(\omega)=\frac{1}{\pi}\int_{-\infty}^{\infty}\int_{-\infty}^{\infty}x_1x_2\psi(x_1,x_2,\omega)\,dx_1\,dx_2. \tag{27.17}$$

This equation gives $G(\omega)$ as the second moment of the distribution of some random process; this distribution $\psi(x_1, x_2, \omega)$ is the Fourier transform for the distribution $\varphi(x_1, x_2, \tau)$ of the original process.

We can obtain one additional relation, proceeding from the definition of the running spectrum of a random process

$$S_t(\omega)=\int_{-\infty}^{t}\xi(x)\,e^{-j\omega x}\,dx. \tag{27.18}$$

Let us introduce the random function $\xi_t(x)$, defined as follows:

$$\xi_t(x) = \begin{cases} \xi(x) & \text{for } x < t, \\ 0 & \text{for } x > t. \end{cases}$$

With this definition we can rewrite (27.18) in the form

$$S_t(\omega) = \int_{-\infty}^{\infty} \xi_t(x) e^{-j\omega x} dx. \tag{27.19}$$

The inverse transform for (27.19) becomes

$$\xi_t(x) = \frac{1}{2\pi} \int_{-\infty}^{\infty} S_t(\omega) e^{j\omega x} d\omega. \tag{27.20}$$

Let us multiply both sides of this equation by $\xi_t(x-\tau)\,dx$ and integrate within infinite limits:

$$\int_{-\infty}^{\infty} \xi_t(x) \xi_t(x-\tau)\, dx = \frac{1}{2\pi} \int_{-\infty}^{\infty} \xi_t(x-\tau)\, dx \int_{-\infty}^{\infty} S_t(\omega) e^{j\omega x} d\omega. \tag{27.21}$$

Let us carry out the integration first with respect to _x_ on the right-hand side. The integral

$$\int_{-\infty}^{\infty} \xi_t(x- \,\,) e^{j\omega x} dx = e^{j\omega\tau} \int_{-\infty}^{\infty} \xi_t(x) e^{j\omega x} dx = e^{j\omega\tau} S_t^*(\omega)$$

is expressed in terms of the complex conjugate of the running spectrum. If now we return to the function $\xi(x)$ on the left-hand side of (27.21) on the basis of the definition of $\xi_t(x)$, i.e., replace the upper limit of integration in correspondence, then we obtain

$$\int_{-\infty}^{t} \xi(x) \xi(x-\tau)\, dx = \frac{1}{2\pi} \int_{-\infty}^{\infty} |S_t(\omega)|^2 e^{j\omega\tau} d\omega.$$

Let us differentiate both sides with respect to _t_

$$\xi(t) \xi(t-\tau) = \frac{1}{2\pi} \int_{-\infty}^{\infty} \frac{\partial}{\partial t} |S_t(\omega)|^2 e^{j\omega\tau} d\omega.$$

Taking the averages over the set of both sides of this expression, we obtain

$$M\,[\xi(t),\,\xi(t-\tau)] = B(\tau) = \frac{1}{2\pi}\int_{-\infty}^{\infty} M\left(\frac{\partial}{\partial t}\,|\,S_t(\omega)\,|^2\right) e^{j\omega\tau}\,d\omega. \qquad (27.22)$$

Bringing (27.22) and (27.9) together, we see that

$$G(\omega) = \frac{1}{\pi}\,M\left(\frac{\partial}{\partial t}\,|\,S_t(\omega)\,|^2\right). \qquad (27.23)$$

It thus turns out that the spectrum $G(\omega)$ of a random process is the average over the whole set of the instantaneous power spectrum in the definition of Page [see (6.3)]; a similar definition was applied by Lampard [24].

We shall now write down the whole series of equations. In order to follow the best systemization, we note that $B(\tau)$ and $G(\omega)$ are related by the pair of Fourier transformations (27.9) and (27.10). These transformations are symmetric with respect to their structure, from which it is possible to conclude that every equation for $B(\tau)$ should correspond to a paired equation for $G(\omega)$. This symmetry becomes evident when we assemble all the equations cited above and group them in the form of a table:

$B(\tau) = \frac{1}{2}\int_{-\infty}^{\infty} G(\omega)\,e^{j\omega\tau}\,d\omega$ (27.9)	$\pi G(\omega) = \int_{-\infty}^{\infty} B(\tau)\,e^{-j\omega\tau}\,d\tau$ (27.10)
$= M\,[\xi(t)\,\xi(t-\tau)]$ (27.4)	$= M\left(\frac{\partial}{\partial t}\left\|S_t(\omega)\right\|^2\right)$ (27.23)
$= \int_{-\infty}^{\infty}\int_{-\infty}^{\infty} x_1x_2\varphi(x_1,\,x_2,\,\tau)\,dx_1\,dx_2$ (27.4)	$= \int_{-\infty}^{\infty}\int_{-\infty}^{\infty} x_1x_2\psi(x_1,\,x_2,\,\omega)\,dx_1\,dx_2$ (27.17)
$= \lim_{T\to\infty}\frac{1}{T}\int_{0}^{T}\xi(t)\,\xi(t-\tau)\,dt$ (27.6)	$= \lim_{T\to\infty}\frac{1}{T}\,\|\,S_T(\omega)\,\|^2$ (27.16)

Further, we note the following in reference to these equations: the last row of the table contains the relations (27.6) and (27.16), which were obtained on the basis of the ergodicity of the stationary process and which represent time averages. These equations can serve as a basis for constructing apparatus for measuring $B(\tau)$ and $G(\omega)$ – correlation meters and analyzers.

But these equations cannot be used directly for the computation of $B(\tau)$ and $G(\omega)$, since they are time averages for some realization. Consequently, in order to carry out the computations from these equations it is necessary to expand the analytic expression for the realization in the form of a certain (nonrandom) function, defined over the entire time axis. In reality a random process is not specified mathematically by one of its realizations, but by some sort of distribution; thus both the correlation function $B(\tau)$ and the spectrum $G(\omega)$ are computed with the inevitable averaging over the set. On the other hand, experimentally we most frequently have to deal precisely with a single realization of a random process. The technique of calculating spectra is demonstrated in the examples of the next section.

For the spectra of random processes we can derive a number of relations analogous to those cited in Section 4. Some of these relations are collated in the following table:

$\eta(t)$	$B_{\eta}(\tau)$	$G_{\eta}(\omega)$
$\xi_1(t)+\xi_2(t)$	$B_1(\tau)+B_2(\tau)$	$G_1(\omega)+G_2(\omega)$
$\xi(at)$	$B(a\tau)$	$G\left(\frac{\omega}{a}\right)$
$\xi(t-t_0)$	$B(\tau)$	$G(\omega)$
$\xi(t)e^{-j\Omega t}$	$B(\tau)e^{j\Omega\tau}$	$G(\omega+\Omega)$
$\xi^{(n)}(t)$	$(-1)^n B^{(2n)}(\tau)$	$\omega^{2n} G(\omega)$
$\xi_1(t)\xi_2(t)$	$B_1(\tau)B_2(\tau)$	$\frac{1}{\pi}\int_{-\infty}^{\infty} G_1(\nu)G_2(\omega-\nu)\,d\nu$

It is assumed that $M\xi = 0$, that ξ_1 and ξ_2 are independent, and that the condition for differentiability is fulfilled whenever demanded.

For the spectra of random processes and their correlation functions the general nature of the connection between the spectral width and the magnitude of the so-called correlation interval holds true, namely,

$$\Delta f \Delta\tau \geqslant \mu, \tag{27.24}$$

where μ is a constant of the order of unity. The correlation interval $\Delta\tau$ expresses the effective duration or extent of the correlation function and can be defined in various ways. General considerations relative to the relation (27.24), valid for any functions connected by a Fourier transformation were presented in Section 12. Here we shall indicate some simple definitions [25], which bring (27.24) to the identity

$$\Delta f \Delta \tau \equiv 1. \tag{27.25}$$

Let us write the expressions for the spectrum and correlation function:

$$G(\omega) = \frac{1}{\pi} \int_{-\infty}^{\infty} B(\tau) \cos \omega\tau \, d\tau,$$

$$B(\tau) = \frac{1}{2} \int_{-\infty}^{\infty} G(\omega) \cos \omega\tau \, d\omega.$$

To a first approximation we shall assume that $\omega = 0$, while in the second approximation $\tau = 0$; we obtain

$$\left.\begin{aligned} G(0) &= \frac{1}{\pi} \int_{-\infty}^{\infty} B(\tau)\, d\tau, \\ B(0) &= \frac{1}{2} \int_{-\infty}^{\infty} G(\omega)\, d\omega. \end{aligned}\right\} \tag{27.26}$$

Let us introduce the definitions

$$\left.\begin{aligned} \Delta\omega &= 2\pi\Delta f = \frac{1}{G(0)} \int_{-\infty}^{\infty} G(\omega)\, d\omega, \\ \Delta\tau &= \frac{1}{B(0)} \int_{-\infty}^{\infty} B(\tau)\, d\tau. \end{aligned}\right\} \tag{27.27}$$

From this, on the basis of (27.26), we immediately obtain (27.25). The meaning of the definitions (27.27) is contained in the fact that the area under the curves for $G(\omega)$ and $B(\tau)$ are equated to the areas of rectangles with the bases $\Delta\omega$ and $\Delta\tau$ and heights $G(0)$ and $B(0)$, respectively.

Let, for example,

$$B(\tau) = e^{-\alpha(\tau)}, \quad G(\omega) = \frac{2}{\pi} \frac{\alpha}{\alpha^2 + \omega^2}.$$

We have

$$\int_{-\infty}^{\infty} B(\tau)\,d\tau = \frac{2}{\alpha}, \qquad B(0) = 1, \qquad \Delta\tau = \frac{\alpha}{2};$$

$$\int_{-\infty}^{\infty} G(\omega)\,d\omega = 2, \qquad G(0) = \frac{2}{\pi\alpha}, \qquad \Delta\omega = \frac{1}{\pi\alpha}, \qquad \Delta f = \frac{2}{\alpha}$$

(see Fig. 45).

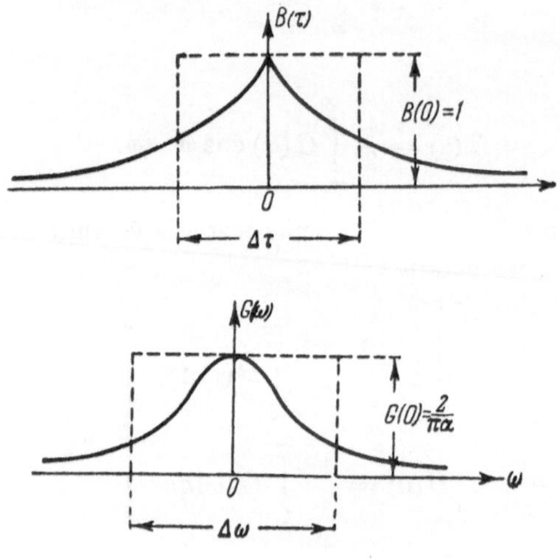

Fig. 45.

The definitions (27.27) also have shortcomings. The first one is contained in the fact that these definitions are inapplicable directly in the case G (0) = 0 [B(0) is never equal to zero]. Among other things, this case is frequently encountered. In particular, we might be interested in a process whose spectrum is contained in the frequency band between ω_1 and ω_2. It is possible, by correspondingly altering the definitions, to represent the graph of the spectrum in this case also by an equivalent rectangle. As for the correlation function, in the case being considered it has an oscillatory character, and it is here that the second shortcoming of the definitions (27.27) appears. It is contained in the fact that, while the function G (ω) is always nonnegative, the function B(τ) can change sign, so that the integral of the function in this case gives a poor criterion for evaluating the correlation interval. These shortcomings are eliminated in definitions that contain any kind of moments of the squares of the functions (see Section 12) or their absolute values, etc.

But in perfecting the definitions in this direction, we lose their simplicity and are deprived of the advantage of applying the identity (27.25).

28. Spectra of Certain Stationary Processes

For the first example we shall consider the stationary random process represented by a random quantity, which can take on with equal probability one of the two values ± a, with an alternation in sign occurring at random times. In Fig. 46 one of the realizations of such a process is illustrated. For the total characteristic of the process it is necessary to specify further a distribution for the points of transition through zero, or, in short, the distribution of zeros. Let this distribution be given in the form of the probability $p(n, \tau)$ that on the interval τ exactly $\underline{n}$ zeros will appear.

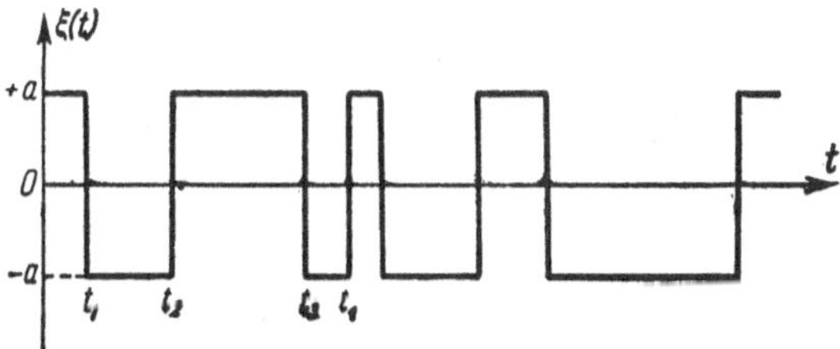

Fig. 46.

Let us first determine the correlation function, applying its basic definition (27.4). In order to do this we must beforehand write an expression for the two-dimensional probability density [i.e., the joint probability density for the random quantities $\xi = \xi(t)$ and $\xi_\tau = \xi(t+\tau)$ to exist in the intervals $(x_1, x_1 + dx_1)$ and $(x_2, x_2 + dx_2)$]. In the present instance the distribution is discrete and has the following properties. If on the interval τ there eixsts an even number of zeros, then $\xi(t)$ and $\xi(t+\tau)$ have the same sign, i.e., with equal probability they are equal to either +a, +a or −a, −a. If now on the interval τ the number of zeros is odd, then $\xi(t)$ and $\xi(t+\tau)$ have opposite signs and are equal to either +a, −a or −a, +a. This can all be written analytically by means of the delta-function as follows:

$$\varphi(x_1, x_2, \tau) =$$

$$= \frac{1}{2}[\delta(x_1 - a, x_2 - a) + \delta(x_1 + a, x_2 + a)] \sum_{k=0}^{\infty} p(2k, \tau) +$$

$$+ \frac{1}{2}[\delta(x_1 - a, x_2 + a) + \delta(x_1 + a, x_2 - a)] \sum_{k=0}^{\infty} p(2k+1, \tau).$$

In this expression

$$\sum_{k=0}^{\infty} p(2k,\tau) = P_e, \quad \sum_{k=0}^{\infty} p(2k+1,\tau) = P_o$$

are the probabilities that an even and odd (respectively) number of zeros will appear on the interval τ.

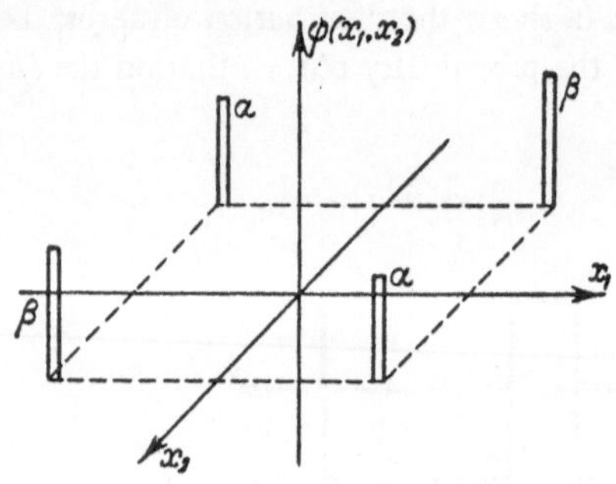

Fig. 47.

For visualization the distribution for the process considered is illustrated in Fig. 47. The two-dimensional delta-functions are arbitrarily represented by columns of a finite height (and volume $\alpha = \frac{1}{2} P_e$ or $\beta = \frac{1}{2} P_o$).

Let us substitute the distribution thus found into the general expression for the correlation function:

$$B(\tau) = \frac{1}{2} P_e \int_{-\infty}^{\infty}\int_{-\infty}^{\infty} x_1 x_2 [\delta(x_1 - a, x_2 - a) + \delta(x_1 + a, x_2 + a)]\, dx_1 dx_2 +$$

$$+ \frac{1}{2} P_o \int_{-\infty}^{\infty}\int_{-\infty}^{\infty} x_1 x_2 [\delta(x_1 - a, x_2 + a) + \delta(x_1 + a, x_2 - a)]\, dx_1\, dx_2.$$

In computing the integrals with the delta-functions it must be borne in mind that by definition this function is

$$\int_{-\infty}^{\infty} f(x)\,\delta(x - x_0)\, dx = f(x_0).$$

Taking this relation into account, after the double integration we obtain

$$B(\tau) = a^2 (P_e - P_o).$$

Let us now consider the special case of the Poisson distribution

$$p(n,\tau)=\frac{(\mu\tau)^n}{n!}e^{-\mu\tau},$$

where μ is the mean frequency of zeros. In this special case we have

$$B(\tau)=a^2e^{-\mu\tau}\left(\sum_{k=0}^{\infty}\frac{(\mu\tau)^{2k}}{(2k)!}-\sum_{k=0}^{\infty}\frac{(\mu\tau)^{2k+1}}{(2k+1)!}\right)$$
$$=a^2e^{-\mu\tau}\sum_{k=0}^{\infty}(-1)^k\frac{(\mu\tau)^k}{k!}$$

or

$$B(\tau)=a^2e^{-2\mu\tau}. \qquad (28.1)$$

From here we then find the spectrum [according to (27.12)]:

$$G(\omega)=\frac{2}{\pi}a^2\int_0^{\infty}e^{-2\mu\tau}\cos\omega\tau\,d\tau=\frac{2}{\pi}a^2\frac{2\mu}{4\mu^2+\omega^2}. \qquad (28.2)$$

We obtained the required result, relying on the fundamental definitions of $B(\tau)$ and $G(\omega)$. However, the same result can be obtained much more rapidly, by finding a way around the formation and integration of the two-dimensional distribution – more precisely, substituting verbal reasoning for an appropriate part of the computation. The argument of Rice, for example, would be applicable here ([30], p. 118 [of the Russian translation]).

Consider the product $\xi\xi_\tau$. It is either equal to a^2 ro to $-a^2$, depending on whether there are an even or odd number of zeros on the interval τ. The mean value of the product $\xi\xi_\tau$ will thus be equal to a^2p_e plus $(-a^2)p_0$, where p_e and p_0 are the probabilities that an even and odd (respectively) number of zeros will appear on the interval τ. Thus

$$B(\tau)=a^2(p_e-p_0)$$

(in the original version the reasoning was only slightly longer).

For the next example we shall take a stationary process characterized by the fact that on the interval T_k between two randomly situated points the process has a random, but fixed value ξ_k ([6], pp. 372-376). The graph of a possible realization of such a process is shown in Fig. 48. In order to determine the correlation function an argument differing from the preceding one is required.

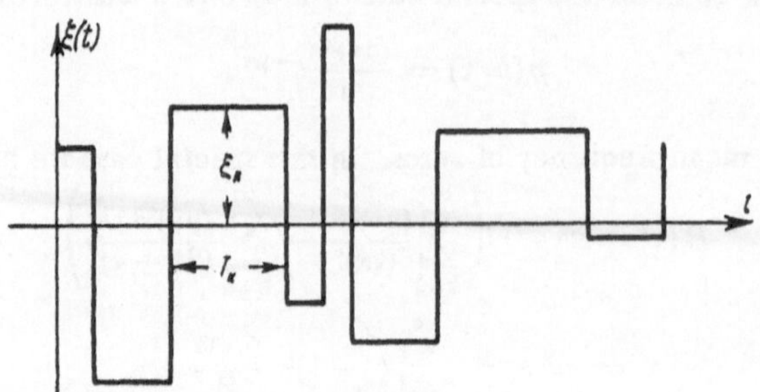

Fig. 48.

The product of the two values $\xi(t)$ and $\xi(t+\tau)$ will be different, depending on whether the times t and $t+\tau$ lie in the same interval T_k or in different ones. In the first instance

$$\xi(t)\xi(t+\tau)=\xi_k^2.$$

In the second

$$\xi(t)\xi(t+\tau)=\xi_i\xi_k$$

and in order to obtain the correlation function it is necessary to add the averages over the set of the values of these products, the latter being multiplied by the corresponding probabilities. Thus we must first of all find the probability that a segment of length τ will be fully contained within the interval T_k. This probability is equal to the probability that over a segment of length τ picked at random from the time axis not a single end point between the intervals T_k, i.e., not a single zero, will appear. This probability is determined as soon as the distribution $p(n,\tau)$ is given; the probability that we are seeking is $p(0,\tau)$. Thus, in the case of the Poisson distribution

$$p(0,\tau)=e^{-\mu\tau}=p(\tau).$$

The probability that the segment τ will not be contained within the interval T_k (i.e., that t and $t+\tau$ will be found in different intervals, for example, T_k and T_i) is obviously equal to $1-p(\tau)$. Therefore, for the correlation function we have

$$B(\tau)=M(\xi_k^2)\,p(\tau)+M(\xi_i\xi_k)\,[1-p(\tau)].$$

Considering that ξ_i and ξ_k are independent and distributed identically, that $M(\xi)=0$ and that, consequently, $M(\xi_i\xi_k)=0$, we obtain

$$B(\tau) = M(\xi^2)\, p(\tau),$$

while for the case of the Poisson distribution

$$B(\tau) = M(\xi^2)\, e^{-\mu\tau}. \tag{28.3}$$

From this the spectrum follows directly:

$$G(\omega) = \frac{2}{\pi} M(\xi^2) \frac{\mu}{\mu^2 + \omega^2}. \tag{28.4}$$

It is interesting to note the difference between Eqs. (28.1) and (28.3). It would seem that the process in Fig. 46 discussed earlier was a special case of the process in Fig. 48 and that it should be possible to obtain an expression for the correlation function for the process in Fig. 46 from Eq. (28.3), assuming therein that

$$M(\xi^2) = a^2.$$

As we can see, this is not the case. The difference is explained by the fact that for the process in Fig. 46, which alternately takes on the values $\pm a$, it is impossible to just let

$$M(\xi_i \xi_k) = 0,$$

as is done for the more general process in Fig. 48. The product $\xi_i \xi_k$ alternately takes on the values $\pm a^2$, so that there is a dependence on the distance between intervals (i.e., on the difference $k - i$), i.e., in the final analysis, on τ. This dependence can be taken into account when the probabilities for even or odd numbers of zeros are chosen, which then leads to a reduction in the correlation function according to the law $e^{-2\mu\tau}$, rather than $e^{-\mu\tau}$.

In the preceding examples the spectrum was computed as the Fourier transform for correlation function. Now we shall demonstrate how the computation is made when, as the basis, the following equation is selected:

$$G(\omega) = \frac{1}{\pi} \lim_{T \to \infty} \frac{1}{T} |S_T(\omega)|^2 .$$

As an example we shall take the process $\pm a$ with a change of sign at random times (Fig. 45) (see [10], pp. 48-49 [18], pp. 194-196). The expression for a realization can be written in the form

$$\xi(t) = a \sum_{k=-\infty}^{\infty} (-1)^k \sigma(t - t_k).$$

For the running spectrum we have

$$S_T(\omega) = a \int_0^T \left[\sum_{k=0}^{\infty} (-1)^k \sigma(t - t_k) \right] e^{-j\omega t}\, dt.$$

Let us change the order of summation and integration, and, at the same time, substitute the appropriate limits, assuming

$$T = nT_{\text{av}},$$

where $T_{\text{av}} = \frac{1}{\mu}$ is the average interval between neighboring zeros. Then we obtain

$$S_T(\omega) = a \sum_{k=1}^{n} (-1)^k \int_{-\infty}^{\infty} \sigma(t - t_k)\, e^{-j\omega t}\, dt.$$

The integral expresses the spectrum of a delayed unit function (see Section 11), so that

$$S_T(\omega) = \frac{a^2}{j\omega} \sum_{k=1}^{n} (-1)^k e^{-j\omega t_k}.$$

In order to obtain the square of the modulus we multiply by the conjugate quantity

$$|S_T(\omega)|^2 = S_T(\omega)\, S_T^*(\omega) = \frac{a^2}{\omega^2} \sum_{k=1}^{n} \sum_{l=1}^{n} (-1)^{k+l} e^{-j\omega(t_k - t_l)}.$$

Thus the expression for the spectrum of the random process has the form

$$G(\omega) = \frac{a^2\mu}{\pi\omega^2} \lim_{n\to\infty} \frac{1}{n} \sum_{k=1}^{n} \sum_{l=1}^{n} (-1)^{k+l} e^{-j\omega(t_k - t_l)}.$$

Let us introduce $m = k - l$ and rewrite the double summation in the form

$$\frac{1}{n}\sum_{k=1}^{n}\sum_{l=1}^{n}(-1)^{k+l}e^{-j\omega(t_k-t_l)} =$$

$$= \frac{1}{n}\sum_{m=-(n-1)}^{n-1}(n-m)(-1)^m e^{-j\omega(t_{l+m}-t_l)}.$$

(Essentially, there are n^2 components in all. Of these, for $\underline{n}$ of the components $m = 0$; for $n-1$ components $m = 1$, and for the same number $m = -1$; for $n-2$ components $m = 2$ and for $n-2$ components $m = -2$, and so forth.)

Let us consider the exponential factor separately. It can be transformed as follows:

$$e^{-j\omega(t_{l+m}-t_l)} = e^{-j\omega(t_{l+m}-t_{l+m-1})}e^{-j\omega(t_{l+m-1}-t_{l+m-2})}\ldots$$

$$\ldots e^{-j\omega(t_{l+1}-t_l)} = e^{-j\omega T_1}e^{-j\omega T_2}\ldots e^{-j\omega T_m} = \prod_{i=1}^{m} e^{-j\omega T_i},$$

where T_i is the sequence of intervals between zeros over the interval $t_k - t_l = t_{l+m} - t_l$.

Let us now insert the factor $\frac{1}{n}$ after the summation sign, pass to the limit, and write the sums for positive and negative values of $\underline{m}$ separately. We obtain

$$G(\omega) = \frac{a^2\mu}{\pi\omega^2}\left[1 + \sum_{m=1}^{\infty}(-1)^m\prod_{i=1}^{m}e^{-j\omega T_i} + \sum_{m=1}^{\infty}(-1)^m\prod_{i=1}^{m}e^{+j\omega T_i}\right].$$

In order to complete the computation we should know the sequence T_i, which defines the given realization. This sequence is not given; the density of the distribution $\varphi(T)$ of the intervals T_i is given. Therefore, at the given stage we resort to averaging over the set.

The problem consists in determining the mean value of the sum

$$\sum_{m=1}^{\infty}(-1)^m\prod_{i=1}^{m}e^{-j\omega T_i}.$$

Let us assume that the intervals T_i are statistically independent. Then the quantities $e^{-j\omega T_i}$ are also independent and the average of the product is equal to the product of the averages. We can write

$$M\left[\sum_{m=1}^{\infty}(-1)^m \prod_{i=1}^{m} e^{-j\omega T_i}\right] = \sum_{m=1}^{\infty}(-1)^m \prod_{i=1}^{m} M\left(e^{-j\omega T_i}\right).$$

It is assumed that all the intervals T_i have the same distribution, expressed by the density $\varphi(T)$. Then

$$M\left(e^{-j\omega T_i}\right) = \int_0^{\infty} \varphi(T)\, e^{-j\omega T}\, dT = \chi(\omega),$$

i.e., the mean value of $e^{-j\omega T}$ is the Fourier transform for the probability density $\varphi(T)$. The function $\chi(\omega)$ is called the characteristic function. Now we can write

$$\sum_{m=1}^{\infty}(-1)^m \prod_{i=1}^{m} M\left(e^{-j\omega T_i}\right) = \sum_{m=1}^{\infty}(-1)^m \chi^m(\omega) = -\frac{\chi(\omega)}{1+\chi(\omega)}.$$

The second sum gives the conjugate quantity. Adding the complex conjugates, we obtain twice the real part, so that, finally,

$$G(\omega) = \frac{a^2\mu}{\pi\omega^2}\left[1 - 2\,\mathrm{Re}\left(\frac{\chi(\omega)}{1+\chi(\omega)}\right)\right].$$

Let us determine the spectrum for the special case of the Poisson distribution of zeros:

$$\varphi(T) = \mu e^{-\mu T}.$$

We now determine the characteristic function

$$\chi(\omega) = \int_0^{\infty} \varphi(T)\, e^{-j\omega T}\, dT = \mu \int_0^{\infty} e^{-(\mu + j\omega)T}\, dT = \frac{\mu}{\mu + j\omega}.$$

Furthermore,

$$\frac{\chi(\omega)}{1+\chi(\omega)} = \frac{\mu}{2\mu + j\omega},$$

and, as a result,

$$G(\omega)=\frac{2}{\pi}a^2\frac{2\mu}{4\mu^2+\omega^2},$$

which agrees with (28.2).

29. Spectrum Concept as Applied to Nonstationary Processes

In the preceding we occupied ourselves only with ergodic stationary processes. However, in practice one must deal with nonstationary processes, and it is without a doubt that such processes are encountered quite often and are of the utmost importance. Many forms of communication signals fall in the category of nonstationary processes. Thus, e.g., an ordinary modulated oscillation

$$\eta(t)=\xi(t)\cos\omega t,$$

where ξ (t) is a stationary random process, is a nonstationary process, this being immediately evident from the fact that the uniform distribution is proportional to cos ωt, i.e., depends on the time. With this understanding it remains necessary to consider the applicability of spectral representations to the class of nonstationary processes of interest here. The problem reduces, of course, to the reexamination and development of the definitions that we have made use of until now.

As we have seen (see Section 27), the computation of the spectrum of a random process is directly connected with averaging operations. For ergodic stationary processes the average over the set and over time are equal to one another with a probability of unity. Therefore, it is of little consequence what kind of averaging technique we employ. The choice of averaging is determined by the way in which we specify the random process. Under experimental conditions we have to deal with a single realization of a random process ordinarily; under these conditions we resort to time averaging. However, under the conditions of theoretical investigation of a random process the probability distribution is given, and in this case we take the average over the set.

Let us consider a random process as a function $\xi(\vartheta, t)$ of two variables: the chance parameter ϑ and the continuously passing time t. Fixing on some instant $t = t_0$, we obtain the random quantity $\xi(\vartheta, t_0)$, the distribution of which depends in the general case on t_0. If instead we attach some value to the random parameter ϑ, we obtain one of the possible realizations of the random process, where this realization is represented by a nonrandom time function.

The operation of averaging is essentially the operation of integration with some kind of weighting. Thus, the time average is found from the expression

$$\overline{\xi(t)} = \lim_{T \to \infty} \frac{1}{T} \int_{-\frac{T}{2}}^{\frac{T}{2}} \xi(t)\, dt.$$

The average over the set is computed as follows:

$$M(\xi) = \int_{-\infty}^{\infty} x\varphi(x)\, dx.$$

However, if we are dealing with a function of two variables, then in the general case the average taken over one variable gives a quantity that depends on the second variable, and vice versa. Applied to random processes this means that the average over the set in general will depend on the time, while the average over the time will form a random set. An ergodic stationary process is characterized by the fact that the averages over time and the set are equal to one another, but from this it follows directly that they do not depend on a second variable (since the functions of equal arguments can be equal to one another only in the case when these functions are constants, i.e., quantities that do not depend on any argument).

Thus the average for a nonstationary process, considered as a function of two variables, can be defined as the result of a two-fold averaging — over the set and over time. The order of averaging is immaterial, since the whole thing reduces simply to a change in the order of integration. Thus, by taking the average over the set for a nonstationary process, we obtain a quantity that depends on the time; in order to eliminate this dependence we must take a second average, this time over the time. Averaging first over the time, we obtain a random set; we are required to take a second average over this set.

In order to complete the picture we should develop the problem of determining for which class of nonstationary processes the average in the sense indicated, first of all, exists, and, secondly, is not deprived of the physical meaning that we ordinarily ascribe to an averaged quantity. Without undertaking such an investigation here, we merely note that as examples in the discussions that follow only those nonstationary processes for which the correlation function and the spectrum, computed by means of a double average, retain their usual sense are considered.[1] In order to avoid misunderstanding we shall designate the correlation function and the spectrum of nonstationary processes, where these are obtained by taking a double average, the mean correlation function and the mean spectrum.

Let us now show that the mean correlation function and spectrum, as before, are interconnected by a pair of Fourier transforms. Let us initially assume the following two definitions:

$$B(\tau, t) = M[\xi(t)\xi(t-\tau)], \qquad G(\omega) = \frac{1}{\pi} M\left(\frac{\partial}{\partial t} |S_t(\omega)|^2\right).$$

Let us write an expression for the running spectrum:

$$S_t(\omega) = \int_{-\infty}^{t} \xi(t_1) e^{-j\omega t_1} dt_1$$

and multiply by the conjugate quantity

$$S_t(\omega) S_t^*(\omega) = |S_t(\omega)|^2 = \int_{-\infty}^{t}\int_{-\infty}^{t} \xi(t_1)\xi(t_2) e^{-j\omega(t_1-t_2)} dt_1 dt_2.$$

We now introduce

$$\tau = t_1 - t_2$$

and divide the region of integration into two parts, as shown in Fig. 49. This gives

$$|S_t(\omega)|^2 = \int_{-\infty}^{t} dt_1 \int_{-\infty}^{0} \xi(t_1)\xi(t_1-\tau) e^{-j\omega\tau} d\tau +$$

$$+ \int_{-\infty}^{t} dt_2 \int_{0}^{\infty} \xi(t_2)\xi(t_2+\tau) e^{-j\omega\tau} d\tau.$$

[1]Bunimovich [3] applies the double average with nonstationary processes of the form $\eta(t) = \psi[\xi(t), f(t)]$ in mind, where ψ is the symbol for some function, $\xi(t)$ is a random process, $f(t)$ is a periodic or quasi-periodic function. All the examples of the following section belong to this class.

Let us now average both sides of this equation over the set, noting that

$$M[\xi(t_1)\xi(t_1-\tau)] = B(t_1, t_1-\tau) = B(\tau, t_1) = M[\xi(t_1)\xi(t_2)] = \\ = M[\xi(t_2)\xi(t_2+\tau)].$$

We obtain

$$M|S_t(\omega)|^2 = 2\int_{-\infty}^{t} dt_1 \int_0^{\infty} B(\tau, t_1)\cos\omega\tau\, d\tau.$$

Differentiating with respect to t and substituting the result into the initial definition of the spectrum, we find

$$G(\omega, t) = \frac{2}{\pi}\int_0^{\infty} B(\tau, t)\cos\omega\tau\, d\tau.$$

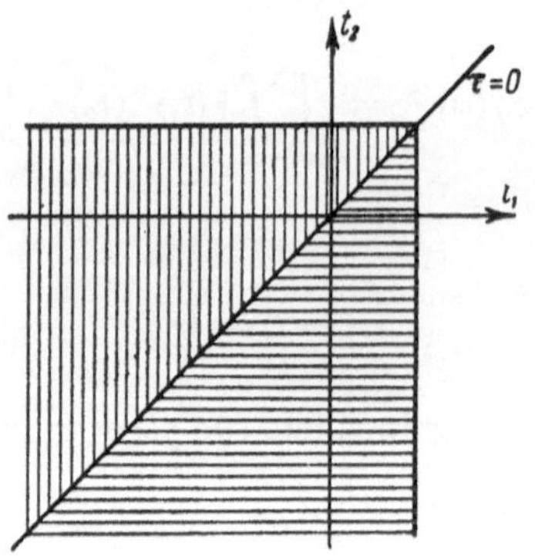

Fig. 49.

From here we take the time average and obtain the connection between the mean spectrum and mean correlation function:

$$\overline{G(\omega, t)} = \frac{2}{\pi}\int_0^{\infty} \overline{B(\tau, t)}\cos\omega\tau\, d\tau.$$

The inverse transformation is also valid:

$$\overline{B(\tau, t)} = \int_0^\infty \overline{G(\omega, t)} \cos \omega\tau \, d\omega$$

[compare with Eqs. (27.11) and (27.12)].

One conclusion following from the above is that we can apply the equations of Section 27 for computing the correlation functions and spectra and nonstationary processes, by taking the second average in each case. From the point of view of the computations it is essential that the second averaging operation can be done at any stage of the computation, as shown in the examples of the next section.

30. Spectra of Some Nonstationary Processes

The distinctive indication of a nonstationary (in the broad sense) process is the fact that its one- and two-dimensional distributions or, in correspondence, the first and second moments, depend on the time. In other words, the quality of being nonstationary is expressed in the fact that the mean value of either the dispersion or the correlation function is a function of time. We shall apply this criterion in the examples considered below.

For the first example we shall consider the random process characterized by the random quantity assuming one of the two equally probable values ± a; the change of sign takes place only at the fixed instants

$$t = t_k = kT_0.$$

An instance of the process described, for example, is a telegraph signal transmitted by a uniform code. One of the realizations is shown in Fig. 50.

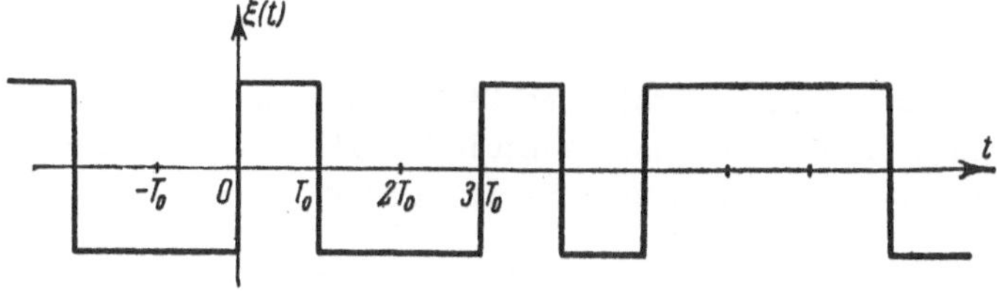

Fig. 50.

Let us write the expression for the probability distribution. The one-dimensional distribution consists of two probabilities:

$$p(a)=p(-a)=\frac{1}{2}.$$

The expression for the one-dimensional probability density can be written by means of the delta function:

$$\varphi(x)=\frac{1}{2}[\delta(x-a)+\delta(x+a)].$$

The one-dimensional distribution, as we can see, does not depend on the time. Let us now consider the distribution for the two-dimensional quantity ξ, ξ_τ, which is a set of two values of a random process separated by τ. The two-dimensional distribution turns out to be time dependent. In fact, if

$$t+\tau<T_0,$$

then the times t and $t+\tau$ are contained within the same interval T_0 and, consequently, ξ and ξ_τ have the same sign (Fig. 51). Therefore,

$$\varphi_1(x_1, x_2)=\frac{1}{2}[\delta(x_1-a, x_2-a)+\delta(x_1+a, x_2+a)] \qquad t+\tau<T_0.$$

If now

$$t+\tau<T_0,$$

then ξ and ξ_τ can, with equal probability, have any sign and, to the point,

$$\varphi_2(x_1, x_2)=\frac{1}{4}[\delta(x_1-a, x_2-a)+\delta(x_1+a, x_2+a)+$$
$$+\delta(x_1-a, x_2+a)+\delta(x_1+a, x_2-a)] \quad t+\tau>T_0.$$

Therefore, the two-dimensional probability density has different expressions on different intervals, which means that it depends not only on τ, but also on t. Thus the process considered is nonstationary.

As pointed out above, the averaging can be done at any stage of the computation. Making use of this fact, let us immediately take the average over the time of the two-dimensional distribution itself. We have

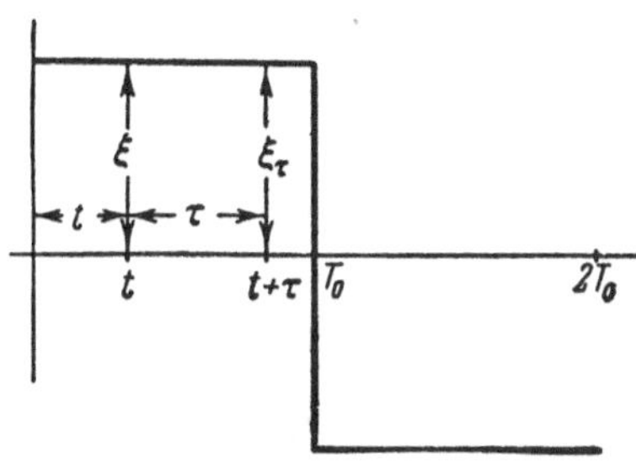

Fig. 51.

$$\varphi(x_1, x_2, \tau, t) = \begin{cases} \varphi_1(x_1, x_2) & \text{for} \quad t+\tau < T_0, \\ \varphi_2(x_1, x_2) & \text{for} \quad t+\tau > T_0. \end{cases}$$

Let us rewrite the inequality

$$t+\tau < T_0$$

in the form

$$0 < t < T_0 - \tau.$$

Averaging the distribution over the interval T_0, we obtain

$$\overline{\varphi(x_1, x_2, \tau, t)} = \frac{1}{T_0} \int_0^{T_0-\tau} \varphi_1(x_1, x_2)\, dt = \left(1 - \frac{\tau}{T_0}\right) \varphi_1(x_1, x_2)$$

for τ, when lying in the interval $0, T_0$. When

$$t+\tau > T_0,$$

in other words, when

$$T_0 - \tau < t < \infty$$

we obtain, averaging over the interval $T \to \infty$,

$$\overline{\varphi(x_1, x_2, \tau, t)} = \varphi_2(x_1, x_2)$$

for τ in the interval T_0, ∞. Thus, for the averaged distribution we can write

$$\begin{aligned}\varphi(x_1, x_2, \tau) &= \overline{\varphi(x_1, x_2, \tau, t)} = \\ &= \frac{1}{2}[\delta(x_1 - a, x_2 - a) + \\ &+ \delta(x_1 + a, x_2 + a)]\left(1 - \frac{\tau}{T_0}\right) \quad \text{for } 0 < \tau < T_0, \\ &= \frac{1}{4}[\delta(x_1 - a, x_2 - a) + \delta(x_1 + a, x_2 + a) + \\ &+ \delta(x_1 - a, x_2 + a) + \delta(x_1 + a, x_2 - a)] \\ &\qquad \text{for } T_0 < \tau < \infty.\end{aligned}$$

Substituting the averaged distribution into the equation

$$\overline{B(\tau, t)} = \int_{-\infty}^{\infty}\int_{-\infty}^{\infty} x_1 x_2 \overline{\varphi(x_1, x_2, \tau, t)}\, dx_1\, dx_2,$$

we find the mean correlation function

$$\overline{B(\tau, t)} = \begin{cases} a^2\left(1 - \frac{\tau}{T_0}\right) & 0 < \tau < T_0, \\ 0 & T_0 < \tau < \infty, \end{cases}$$

then the mean spectrum

$$\begin{aligned}\overline{G(\omega, t)} &= \frac{2}{\pi}\int_0^{\infty} \overline{B(\tau, t)} \cos\omega\tau\, d\tau = \frac{2}{\pi} a^2 \int_0^{T_0}\left(1 - \frac{\tau}{T_0}\right)\cos\omega\tau\, d\tau = \\ &= \frac{1}{\pi} a^2 T_0 \left(\frac{\sin\omega\frac{T_0}{2}}{\omega\frac{T_0}{2}}\right)^2.\end{aligned} \tag{30.1}$$

For the second example we shall take a process somewhat general in form: the time axis is divided as before into equal intervals T_0; in each interval the random quantity assumes the independent values ξ_k, which are determined by the probability density $\varphi(x)$ (Fig. 52). For the two-dimensional probability density we have the following expressions:

$$\varphi(x_1, x_2, \tau, t) = \begin{cases} \varphi(x_1)\delta(x_1 - x_2) & t + \tau < T_0, \\ \varphi(x_1)\,\varphi(x_2) & t + \tau > T_0. \end{cases}$$

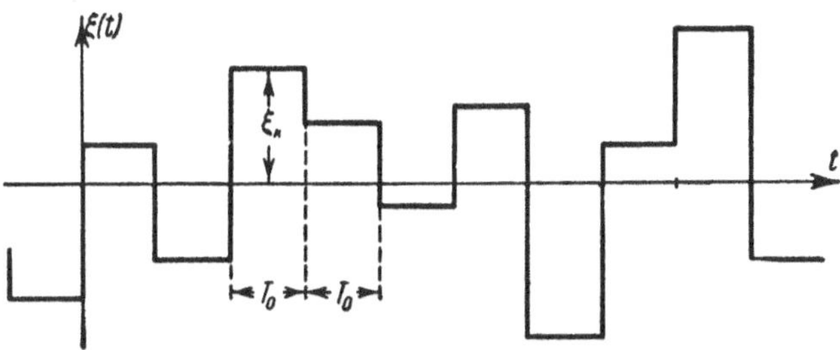

Fig. 52.

If $M(\xi) = 0$, after taking the average we find the correlation function and spectrum:

$$\overline{B(\tau, t)} = \begin{cases} M(\xi^2)\left(1 - \frac{\tau}{T_0}\right) & 0 < \tau < T_0, \\ 0 & T_0 < \tau < \infty, \end{cases} \tag{30.2}$$

$$\overline{G(\omega, t)} = \frac{1}{\pi} T_0 M(\xi^2) \left(\frac{\sin \omega \frac{T_0}{2}}{\omega \frac{T_0}{2}} \right)^2 .$$

The process considered can be represented as a succession of rectangular pulses with a porosity of unity and with amplitude modulation. It must be noted that the spectra of the different forms of pulse modulation are more conveniently found by proceeding from the running spectrum, similar to the pattern shown in Section 28.

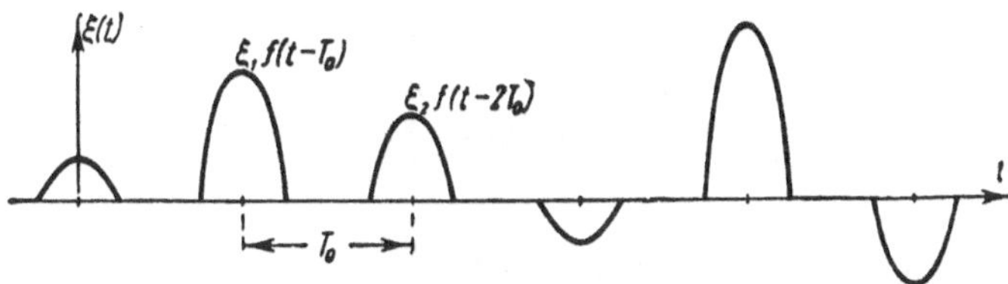

Fig. 53.

Let us consider the general case of amplitude-pulse modulation. The process is represented by a sequence of pulses of arbitrary form $f(t)$, modulated in amplitude, i.e., multiplied by the random factor ξ_k(Fig. 53). Such a process can be written in the form

$$\eta(t)=\sum_{k=-\infty}^{\infty}\xi_k f(t-kT_0).$$

Let us write the expression for the running spectrum:

$$S_T(\omega)=\int_0^T\left[\sum_{k=-\infty}^{\infty}\xi_k f(t-kT_0)\right]e^{-j\omega t}\,dt.$$

Let us change the order of operation with a simultaneous replacement of the limits:

$$S_T(\omega)=\sum_{k=1}^{n}\xi_k\int_{-\infty}^{\infty}f(t-kT_0)\,e^{-j\omega t}\,dt=A(\omega)\sum_{k=1}^{n}\xi_k e^{-jk\omega T_0},$$

where

$$A(\omega)=\int_{-\infty}^{\infty}f(t)\,e^{-j\omega t}\,dt$$

is the spectrum of a single pulse represented by the function $f(t)$. Let us multiply $S_T(\omega)$ by the conjugate quantity

$$S_T(\omega)\,S_T^*(\omega)=|S_T(\omega)|^2=|A(\omega)|^2\sum_{k=1}^{n}\sum_{l=1}^{n}\xi_k\xi_l\,e^{-j(k-l)\omega T_0}.$$

We still must divide by $T = nT_0$, pass to the limit for $T\to\infty$, and take the average over the set:

$$\overline{G(\omega,t)}=\frac{1}{\pi T_0}M\left\{|A(\omega)|^2\lim_{n\to\infty}\frac{1}{n}\sum_{k=1}^{n}\sum_{l=1}^{n}\xi_k\xi_l\;e^{-j(k-l)\omega T_0}\right\}.$$

We then have $\underline{n}$ terms, for which $k = l$; after summing and dividing by $\underline{n}$ these terms will give in the mean $M(\xi^2)$. The mean value of the product $\xi_k\,\xi_l$ (on the assumption that ξ_k and ξ_l are independent) is equal to the product of the means, i.e., $(M\xi)^2$; this quantity can be taken out and put in front of the summation sign. Introducing $m = k - l$, after a few transformations similar to those made in Section 28, we obtain

$$\overline{G(\omega,t)} = \frac{1}{\pi T_0} \mid A(\omega) \mid^2 \left\{ M(\xi^2) - (M\xi)^2 - (M\xi)^2 \sum_{-\infty}^{\infty} e^{-jm\omega T_0} \right\}$$

[here the component $-(M\xi)^2$ is added so that the summation will include this term, which corresponds to the value m = 0]. The summation on the right-hand side is written in the form of a Fourier series of some periodic function of frequency. It is easily seen that this function is nothing other than a delta-function sequence of the form

$$F(\omega) = \omega_0 \sum_{m=1}^{\infty} \delta(\omega - m\omega_0),$$

of which we are easily convinced after expanding the function $F(\omega)$ in a Fourier series. The function $F(\omega)$ represents the harmonic line spectrum; $\omega_0 = \frac{2\pi}{T_0}$ is the fundamental frequency of the spectrum, which is equal to the pulse repetition rate. Thus $G(\omega)$ contains continuous and discrete parts; the latter is absent when $M\xi = 0$. The final expression for the spectrum is written in the form

$$\overline{G(\omega,t)} = \frac{1}{\pi T_0} \mid A(\omega) \mid^2 [M(\xi^2) - (M\xi)^2 + (M\xi)^2 F(\omega)]. \quad (30.3)$$

Let us apply this equation to the process in Fig. 52, which represents a sequence of rectangular pulses of duration T_0. In this case we have

$$A(\omega) = T_0 \frac{\sin \omega \frac{T_0}{2}}{\omega \frac{T_0}{2}}, \qquad M(\xi) = 0$$

and, consequently,

$$\overline{G(\omega,t)} = \frac{1}{\pi} T_0 M(\xi^2) \left(\frac{\sin \omega \frac{T_0}{2}}{\omega \frac{T_0}{2}} \right)^2$$

is the result obtained earlier by another means.

Now let us go on to the case of phase-modulated pulses. The process is represented by a succession of pulses of the same form and size, having the random displacements ϵ_k in time relative to the fixed equal intervals of time kT_0; in other words, the pulse-repetition period T_0 takes on the random increments ϵ_k. The quantities ϵ_k are assumed to be independent; the process is specified by the one-dimensional distribution $\varphi(\epsilon)$. An illustration is given in Fig. 54 for the example of rectangular pulses. We shall not write the expressions for the two-dimensional distribution; we shall merely note that the product $\xi\xi_\tau$ is equal to a^2 when both the following conditions are satisfied simultaneously:

$$\left.\begin{aligned} kT_0+\varepsilon_k-\tfrac{1}{2}\tau_0<t<kT_0+\varepsilon_k+\tfrac{1}{2}\tau_0,\\ lT_0+\varepsilon_l-\tfrac{1}{2}\tau_0<t+\tau<lT_0+\varepsilon_l+\tfrac{1}{2}\tau_0.\end{aligned}\right\}$$

If it happened that one of these inequalities were not observed; then $\xi\xi_\tau = 0$. From this it is evident that the correlation function depends not only on τ, but on t as well, so that the process considered is nonstationary.

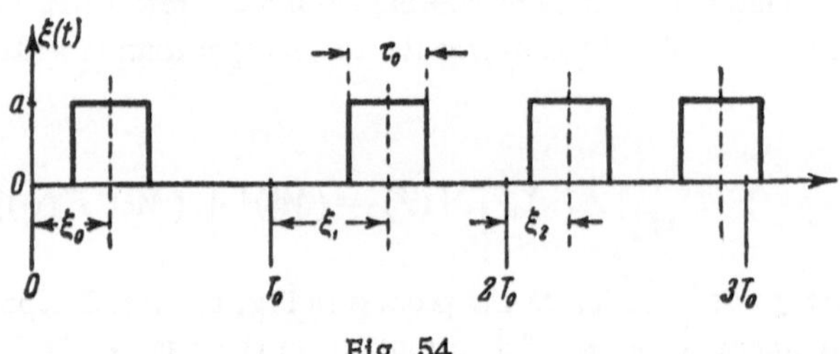

Fig. 54.

We shall compute the spectrum directly for pulses of arbitrary form. The process can be written in the form

$$\xi(t)=\sum_{k=-\infty}^{\infty} f(t-kT_0-\varepsilon_k),$$

where $f(t)$, as before, is the function for the pulse. The running spectrum is expressed as follows:

$$S_T(\omega)=\int_0^T\left[\sum_{k=-\infty}^{\infty} f(t-kT_0-\varepsilon_k)\right]e^{-j\omega t}\,dt$$

or, after changing the order of operation and replacing the limits,

$$S_T(\omega)=\sum_{k=1}^{n}\int_{-\infty}^{\infty} f(t-kT_0-\varepsilon_k)e^{-j\omega t}dt=A(\omega)\sum_{k=1}^{n}e^{-j\omega(kT_0+\varepsilon_k)},$$

where A (ω) is the spectrum of a single pulse. Furthermore,

$$|S_T(\omega)|^2=|A(\omega)|^2\sum_{k=1}^{n}\sum_{l=1}^{n}e^{-j\omega(k-l)T_0}e^{-j\omega\varepsilon_k}e^{j\omega\varepsilon_l}.$$

Averaging over the set gives

$$M|S_T(\omega)|^2=|A(\omega)|^2|\chi(\omega)|^2\sum_{k=1}^{n}\sum_{l=1}^{n}e^{-j\omega(k-l)T_0},$$

where

$$\chi(\omega)=M(e^{-j\omega\varepsilon})=\int_{-\infty}^{\infty}\varphi(\varepsilon)e^{-j\omega\varepsilon}d\varepsilon$$

is the characteristic distribution function for ϵ. Transforming the double summation precisely as was done earlier, we obtain

$$\overline{G(\omega,t)}=\frac{1}{\pi T_0}|A(\omega)|^2\left(1-|\chi(\omega)|^2+|\chi(\omega)|^2\sum_{m=-\infty}^{\infty}e^{-j\omega mT_0}\right)$$

or

$$\overline{G(\omega,t)}=\frac{1}{\pi T_0}|A(\omega)|^2[1-|\chi(\omega)|^2+|\chi(\omega)|^2F(\omega)], \qquad (30.4)$$

where

$$F(\omega)=\omega_0\sum_{m=1}^{\infty}\delta(\omega-m\omega_0).$$

Let us turn now to an investigation of sinusoidal oscillations modulated by a random process in amplitude or phase.

We shall begin with amplitude modulation, which is the simpler case. The random process whose spectrum we must determine is written in the form

$$\eta(t) = \xi(t) \cos \omega_0 t,$$

where ξ (t) is a stationary process. Let us form the product

$$\eta(t)\,\eta(t+\tau) = \frac{1}{2}\xi(t)\,\xi(t+\tau)\,[\cos \omega_0 (2t+\tau) + \cos \omega_0 \tau].$$

Averaging this product over the set, we obtain

$$B_\eta(\tau, t) = \frac{1}{2} B_\xi(\tau)\,[\cos \omega_0 (2t+\tau) + \cos \omega_0 \tau].$$

Let us now take the time average; the first term in the brackets drops out, and we then have

$$\overline{B_\eta(\tau, t)} = \frac{1}{2} B_\xi(\tau) \cos \omega_0 \tau.$$

From this we determine the mean spectrum

$$\overline{G_\eta(\omega,t)} = \frac{2}{\pi}\int_0^\infty \overline{B_\eta(\tau,t)} \cos \omega\tau \, d\tau = \frac{1}{\pi}\int_0^\infty B_\xi(\tau) \cos \omega_0\tau \cos \omega\tau \, d\tau =$$

$$= \frac{1}{2\pi}\int_0^\infty B_\xi(\tau) \cos(\omega_0+\omega)\tau \, d\tau + \frac{1}{2\pi}\int_0^\infty B_\xi(\tau) \cos(\omega_0-\omega)\tau \, d\tau.$$

Thus

$$\overline{G_\eta(\omega,t)} = \frac{1}{4}\,[G_\xi(\omega_0+\omega) + G_\xi(\omega_0-\omega)], \qquad (30.5)$$

i.e., the spectrum of the modulated oscillations consists of two side bands, which repeat the spectrum of the modulating process, and which are distributed symmetrically relative to the carrier frequency ω_0. Thus the ordinary construction of the spectrum of an amplitude modulated oscillation is still preserved in the case when the modulating function is a random process.

Now we wish to consider phase modulation. Here the problem becomes considerably more complex [5].

The task before us is to find the spectrum of a random process of the form

$$\xi(t) = \cos \theta_1(t),$$

where $\theta_1(t)$ is the random process. But since in a sense the phase (argument of the cosine) should in the mean increase in direct proportion to the time, we shall consider a special form of phase modulation, expressed by the relation

$$\xi(t) = \cos[\omega_0 t + \theta(t)]. \tag{30.6}$$

Here $\theta(t)$ can now be treated as a stationary process, which will be our assumption.

First of all it is necessary to show that the process (30.6) is nonstationary. In order to do this we shall compose an expression for the probability density of the two-dimensional random quantity ξ, ξ_τ. We have

$$\left.\begin{aligned} \xi &= \xi(t) = \cos[\omega_0 t + \theta(t)], \\ \xi_\tau &= \xi(t+\tau) = \cos[\omega_0(t+\tau) + \theta(t+\tau)]. \end{aligned}\right\} \tag{30.7}$$

The probability density for the two-dimensional quantity θ, θ_τ is given and is equal to $\varphi_\theta(\vartheta_1, \vartheta_2, \tau)$. The density $\varphi_\xi(x_1, x_2, \tau)$, which we are seeking, should be expressed in terms of φ_θ. The problem reduces to a change of variables according to the well known equation

$$\varphi_\xi(x_1, x_2, \tau) = \varphi_\theta(\vartheta_1, \vartheta_2, \tau)\frac{1}{\dfrac{\partial(x_1, x_2)}{\partial(\vartheta_1, \vartheta_2)}},$$

where

$$\frac{\partial(x_1, x_2)}{\partial(\vartheta_1, \vartheta_2)} = \begin{vmatrix} \dfrac{\partial x_1}{\partial\vartheta_1} & \dfrac{\partial x_1}{\partial\vartheta_2} \\ \dfrac{\partial x_2}{\partial\vartheta_1} & \dfrac{\partial x_2}{\partial\vartheta_2} \end{vmatrix}$$

is the Jacobian of the transformation. In our example x_1 depends only on ϑ_1, while x_2 depends only on ϑ_2. Therefore,

$$\varphi_\xi = \varphi_\theta \frac{1}{\dfrac{dx_1}{d\vartheta_1}\dfrac{dx_2}{d\vartheta_2}}$$

and we have

$$\varphi_\xi(x_1, x_2, \tau, t) = \frac{\varphi_\theta(\vartheta_1, \vartheta_2, \tau)}{\sin[\omega_0 t + \vartheta_1]\sin[\omega_0(t+\tau) + \vartheta_2]}.$$

Thus the probability density φ_ξ depends on the time, and the process (30.6), consequently, is nonstationary.

Let us now compute the spectrum, for which we shall first find the correlation function. We have

$$B_\xi(\tau, t) = M(\xi\xi_\tau) = \int_{-\infty}^{\infty}\int_{-\infty}^{\infty} x_1 x_2 \varphi_\xi(x_1, x_2, \tau, t)\, dx_1\, dx_2.$$

In our example

$$\varphi_\xi(x_1, x_2, \tau, t)\, dx_1\, dx_2 = \varphi_\theta(\vartheta_1, \vartheta_2, \tau)\, d\vartheta_1\, d\vartheta_2.$$

Thus

$$B_\xi(\tau, t) = \int_{-\infty}^{\infty}\int_{-\infty}^{\infty} x_1 x_2 \varphi_\theta(\vartheta_1, \vartheta_2, \tau)\, d\vartheta_1\, d\vartheta_2$$

or, substituting the expressions for x_1 and x_2 in terms of ϑ_1 and ϑ_2 from (30.7),

$$B_\xi(\tau, t) = \int_{-\infty}^{\infty}\int_{-\infty}^{\infty} \cos(\omega_0 t + \vartheta_1)\cos[\omega_0(t+\tau) + \vartheta_2]\, \varphi_\theta(\vartheta_1, \vartheta_2, \tau)\, d\vartheta_1\, d\vartheta_2.$$

Transforming the cosine product, we obtain two terms, one of which drops out upon taking the time average. Therefore,

$$\overline{B_\xi(\tau, t)} = \frac{1}{2}\int_{-\infty}^{\infty}\int_{-\infty}^{\infty} \cos(\omega_0\tau + \vartheta_2 - \vartheta_1)\, \varphi_\theta(\vartheta_1, \vartheta_2, \tau)\, d\vartheta_1\, d\vartheta_2 =$$

$$= \frac{1}{2} M[\cos(\omega_0\tau + \theta_\tau - \theta)] = \frac{1}{2}\cos\omega_0\tau\, M[\cos(\theta - \theta_\tau)] -$$

$$- \frac{1}{2}\sin\omega_0\tau\, M[\sin(\theta - \theta_\tau)].$$

We shall restrict ourselves to the case of a symmetrical distribution of the difference

$$\zeta(\tau, t) = \theta(t) - \theta(t+\tau).$$

In this case the second term drops out, and we obtain

$$\overline{B_\xi(\tau, t)} = \frac{1}{2}\cos\omega_0\tau\, M[\cos\zeta(\tau, t)].$$

Hence we find the spectrum for the process ξ (t):

$$\overline{G_{\xi}(\omega, t)} = \frac{1}{\pi} \int_0^{\infty} \cos\omega_0\tau \cos\omega\tau\, M(\cos\zeta)\, d\tau = \\ = \frac{1}{2\pi} \int_0^{\infty} M(\cos\zeta)\cos(\omega_0 + \omega)\tau\, d\tau + \\ + \frac{1}{2\pi} \int_0^{\infty} M(\cos\zeta)\cos(\omega_0 - \omega)\tau\, d\tau \tag{30.8}$$

or

$$\overline{G_{\xi}(\omega, t)} = \frac{1}{4}\left[G(\omega_0 + \omega) + G(\omega_0 - \omega)\right].$$

Thus in the case of phase modulation the spectrum again consists, as one should expect, of two side bands. The spectrum of each is expressed by the equation

$$G(\omega) = \frac{2}{\pi} \int_0^{\infty} M(\cos\zeta)\cos\omega\tau\, d\tau. \tag{30.9}$$

Let us determine the average value of $\cos\zeta$. Let θ have a normal distribution; in this case the difference

$$\zeta = \theta - \theta_{\tau}$$

also has a normal distribution, which is completely defined by the dispersion

$$D(\zeta) = \sigma^2 = M(\theta - \theta_{\tau})^2 = M\theta^2 + M\theta_{\tau}^2 - 2M\theta\theta_{\tau} = \\ = 2[D(\theta) - B_{\theta}(\tau)] = 2[B_{\theta}(0) - B_{\theta}(\tau)],$$

and we can now determine

$$M(\cos\zeta) = \int_{-\infty}^{\infty} \cos z\varphi(z)\, dz = \frac{2}{\sqrt{2\pi}\,\sigma} \int_0^{\infty} \cos z\, e^{-\frac{z^2}{2\sigma^2}}\, dz = \\ = e^{-\frac{1}{2}\sigma^2} = e^{-[B_{\theta}(0) - B_{\theta}(\tau)]}$$

and the expression for the spectrum

$$G(\omega) = \frac{2}{\pi} \int_0^\infty \cos \omega\tau \, e^{-[B_\theta(0) - B_\theta(\tau)]} \, d\tau. \qquad (30.10)$$

This equation expresses the spectrum of a phase-modulated oscillation in terms of the correlation function of the modulating process, the distribution of which is assumed to be normal.[1] In order to bring the computation to its completion, we shall consider a special case; we shall focus our attention on frequency modulation with a large index.

In the case of frequency modulation the process $\theta(t)$ is not given directly, instead we have the process

$$\nu(t) = \frac{d\theta}{dt},$$

which expresses the instantaneous deviation of the frequency from the mean value ω_0. Therefore, we can write for the frequency-modulated oscillation

$$\xi(t) = \cos\left[\omega_0 t + \int \nu(t)\, dt\right].$$

The spectrum of the process $\xi(t)$ must necessarily be expressed in terms of the characteristic of the process $\nu(t)$. It is useful to apply the Fourier transformation. If

$$B_\nu(\tau) = \int_0^\infty G_\nu(\omega) \cos \omega\tau \, d\omega,$$

then

$$B_\theta(\tau) = \int_0^\infty \frac{1}{\omega^2} G_\nu(\omega) \cos \omega\tau \, d\omega$$

and the quantity σ appearing in the preceding equations can be expressed in terms of the spectrum of the instantaneous frequency deviation $G_\nu(\omega)$ as follows:

$$\frac{1}{2}\sigma^2 = B_\theta(0) - B_\theta(\tau) = \int_0^\infty \frac{1}{\omega^2} G_\nu(\omega)(1 - \cos \omega\tau)\, d\omega.$$

---

[1]A similar equation was obtained by Zadeh [32], using an alternate method.

Let the spectrum G_ν be uniform and bounded above by the frequency Ω, i.e.,

$$G_\nu(\omega) = \begin{cases} g = \text{const} & 0 < \omega < \Omega, \\ 0 & \Omega < \omega < \infty. \end{cases}$$

Then

$$\frac{1}{2}\sigma^2 = g\int_0^\Omega \frac{1-\cos\omega\tau}{\omega^2}\,d\omega.$$

For our purposes it is convenient to write this integral in the following form:

$$\frac{1}{2}\sigma^2 = g\Omega\frac{1}{2}\tau^2 f(\Omega\tau),$$

where

$$f(\Omega\tau) = 1 - \frac{2}{3\cdot 4!}(\Omega\tau)^2 + \frac{2}{5\cdot 6!}(\Omega\tau)^4 - \ldots$$

The product

$$g\Omega = \int_0^\infty G_\nu(\omega)\,d\omega$$

is the mean square of the process $\nu(t)$, i.e., the mean square of the deviation of the frequency from the mean value. We shall denote this quantity by $\Delta\omega^2$. Introducing the notation of the modulation index

$$\beta = \frac{\Delta\omega}{\Omega},$$

we can rewrite the expression for σ^2 in the form

$$\frac{1}{2}\sigma^2 = \frac{1}{2}\beta^2(\Omega\tau)^2 f(\Omega\tau).$$

Now we shall consider an approximation for the case [5] of a large index, i.e., for $\beta \gg 1$.

The case of a large index is characterized by the fact that in the expression for the desired spectrum

$$G(\omega)=\frac{2}{\pi}\int_0^\infty \cos\omega\tau\, e^{-\frac{1}{2}\beta^2(\Omega\tau)^2 f(\Omega\tau)}\,d\tau$$

the exponential factor decreases more rapidly, as β becomes larger. This means that the integrand becomes negligibly small for smaller values of τ, which is the case when β becomes large. Consequently, when β is large we can restrict ourselves to small values of τ. This then serves as the basis for our approximation. In particular, it can be assumed that

$$f(\Omega\tau)\simeq 1$$

with an error not exceeding 0.01, when $\Omega\tau \leq 0.6$. Under this condition we have

$$e^{-\frac{1}{2}\beta^2(\Omega\tau)^2 f(\Omega\tau)}\simeq e^{-\frac{1}{2}\beta^2(\Omega\tau)^2}<\varepsilon,$$

when

$$\beta^2>\frac{2\ln\frac{1}{\varepsilon}}{(\Omega\tau)^2},$$

from which, substituting the greatest value $\Omega\tau = 0.6$ and choosing $\epsilon = 0.01$, we find

$$\beta>5.$$

The considerations leading to the approximation described are easily generalized to the case of an arbitrary spectrum G_ν of the modulating process $\nu(t)$ under the condition that the width Ω of the spectrum G is sufficiently small in comparison with the mean-square frequency deviation $\Delta\omega$. In order to obtain this more general result, it is sufficient to expand $1-\cos\omega\tau$ in a series in the expression for σ^2 and to retain the first term of the series. This gives

$$\frac{1}{2}\sigma^2=\int_0^\infty \frac{1}{\omega^2}G_\nu(\omega)(1-\cos\omega\tau)\,d\tau\simeq$$

$$\simeq\frac{1}{2}\tau^2\int_0^\Omega G_\nu(\omega)\,d\omega=\frac{1}{2}\tau^2\Delta\omega^2=\frac{1}{2}\beta^2(\Omega\tau)^2,$$

i.e., the same as before.

Thus, for a large modulation index, or, what amounts to the same thing, for a large frequency deviation, we have the following approximate expression for the spectrum:

$$G(\omega)=\frac{2}{\pi}\int_0^\infty \cos\omega\tau e^{-\frac{1}{2}\Delta\omega^2\tau^2}\,d\tau=\sqrt{\frac{2}{\pi}}\frac{1}{\Delta\omega}e^{-\frac{1}{2}\frac{\omega^2}{\Delta\omega^2}}. \tag{30.11}$$

Here it should be noted that when the index is large the spectrum of the frequency-modulated oscillation does not depend on the shape and width of the spectrum of the modulating process; the spectrum is determined only by the frequency deviation, where the width of the spectrum of the frequency-modulated oscillation is proportional to the frequency deviation.

31. Remarks on the Analysis of Random Processes

In this section we shall discuss certain problems that arise in the technique of performing spectral analysis on random processes and in connection with the errors that appear in analysis, i.e., in the experimental determination of spectra. For simplicity we shall focus our attention only on stationary ergodic processes.

First of all we note that, as pointed out above, under the conditions of experiment we have to deal ordinarily with one realization of a random process. Therefore, the fundamental objective of spectral analyzers for random processes is the determination of a spectrum containing a time average, for example,

$$G(\omega)=\frac{1}{\pi}\lim_{T\to\infty}\frac{1}{T}|S_T(\omega)|^2, \tag{31.1}$$

where

$$S_T(\omega)=\int_0^T \xi(t)\,e^{-j\omega t}\,dt$$

is the running spectrum of the realization, while the quantity

$$\frac{1}{\pi}|S_T(\omega)|^2$$

expresses the running spectrum of the energy given off by the process $\xi(t)$ during the time T. But the quantity

$$\frac{1}{\pi}\frac{1}{T}|S_T(\omega)|^2 = G_T(\omega)$$

is the spectral density of the power, averaged over the time T, for the process $\xi(t)$. Both $|S_T(\omega)|$ and $G_T(\omega)$ are random quantities. Only in the limit does G_T converge to G, i.e.,

$$G(\omega) = \lim_{T\to\infty} G_T(\omega). \tag{31.2}$$

It is quite clear that under real experimental conditions only a finite time is involved. Therefore, as a result of a single experiment we shall obtain the random quantity $G_T(\omega)$, which, generally speaking, differs from the true value $G(\omega)$. If now the test (of the same duration T) is repeated many times, then a set of values $G_T(\omega)$ is obtained and in the limit we should find the true spectrum, as follows:

$$G(\omega) = MG_T(\omega). \tag{31.3}$$

Let us turn now to a determination of the error of measurement, i.e., the deviation of the measured spectrum from its pure value. First let us find the energy given off by the process $\xi(t)$ during the time T. This energy is equal to

$$E_T = \int_0^T \xi^2(t)\,dt = \frac{1}{\pi}\int_0^\infty |S_T(\omega)|^2\,d\omega$$

or

$$E_T = T\int_0^\infty G_T(\omega)\,d\omega.$$

It is obvious that E_T is also a random quantity. Its mean value is

$$E = ME_T = TM\int_0^\infty G_T(\omega)\,d\omega = T\int_0^\infty G(\omega)\,d\omega.$$

The root-mean-square deviation of E_T from E is then the standard deviation of the measurement. Thus the standard deviation is expressed directly by the dispersion in the quantity E_T:

$$\sigma^2 = DE_T = M\left(E_T^2\right) - M^2\left(E_T\right) = M\left(E_T^2\right) - E^2. \qquad (31.4)$$

We have

$$E_T^2 = \left(T\int\limits_0^\infty G_T(\omega)\,d\omega\right)^2 = T^2\int\limits_0^\infty\int\limits_0^\infty G_T(\nu_1)\,G_T(\nu_2)\,d\nu_1\,d\nu_2.$$

For the mean value of this quantity Rice [29] derived the following expression:

$$M\left(E_T^2\right) = T^2\int\limits_0^\infty\int\limits_0^\infty G(\nu_1)\,G(\nu_2)\left[1+\left(\frac{\sin(\nu_1+\nu_2)\,T/2}{(\nu_1+\nu_2)\,T/2}\right)^2+\right.$$

$$\left.+\left(\frac{\sin(\nu_1-\nu_2)\,T/2}{(\nu_1-\nu_2)\,T/2}\right)^2\right]d\nu_1\,d\nu_2, \qquad (31.5)$$

whence

$$\sigma^2 = T^2\int\limits_0^\infty\int\limits_0^\infty G(\nu_1)\,G(\nu_2)\left[\left(\frac{\sin(\nu_1+\nu_2)\,T/2}{(\nu_1+\nu_2)\,T/2}\right)^2+\right.$$

$$\left.+\left(\frac{\sin(\nu_1-\nu_2)\,T/2}{(\nu_1-\nu_2)\,T/2}\right)^2\right]d\nu_1\,d\nu_2. \qquad (31.6)$$

It is more convenient to normalize the error, i.e., to introduce the relative root-mean-square error (coefficient of variation), defining it as follows:

$$\varepsilon = \frac{\sigma}{E}.$$

Now let us imagine an analyzer in the form of an ideal filter with a transmission coefficient equal to unity in the frequency band from ω_1 to ω_2 and equal to zero outside this inverval. The process of analysis consists in measuring the energy delivered at the output of the filter during the time T. If the analyzer has a high resolving power, then the band

$$\Omega = \omega_2 - \omega_1$$

will be narrow, and we can consider that within the confines of this band the true spectral density will be constant and equal to g_0. In this case

$$E = Tg_0\Omega,$$

and for the standard deviation squared we have

$$\sigma^2 = T^2 g_0^2 \int_{\omega_1}^{\omega_2}\int_{\omega_1}^{\omega_2} \left[\left(\frac{\sin(\nu_1+\nu_2)\,T/2}{(\nu_1+\nu_2)\,T/2}\right)^2 + \left(\frac{\sin(\nu_1-\nu_2)\,T/2}{(\nu_1-\nu_2)\,T/2}\right)^2\right] d\nu_1 d\nu_2.$$

For very small values of T we have (assuming $\sin x/x \approx 1$)

$$\sigma^2 \simeq 2T^2 g_0^2\, \Omega^2$$

and

$$\frac{\sigma}{E} \simeq \sqrt{2};$$

for very large values of T the following asymptotic expression is obtained:

$$\frac{\sigma}{E} \sim \sqrt{\frac{2\pi}{\Omega T}} = \frac{1}{\sqrt{FT}}.$$

It should be noted that an asymptotic expression of the form

$$\frac{\sigma}{E} \sim \frac{A}{\sqrt{\Omega T}}$$

holds not only for an ideal filter, such as we are considering, but also for other types of filters; the constant A depends on the properties of the filter [21].

For the case of a band whose width is small in comparison with the middle frequency, i.e., under the condition

$$\Omega = \omega_2 - \omega_1 \ll \omega_0 = \frac{1}{2}\left(\omega_1 + \omega_2\right)$$

the value of the relative error is treated [29] as a function of the argument FT ($F = f_2 - f_1 = \frac{1}{2\pi}\Omega$) (Rice's computations are very cumbersome, so that we shall not quote them); the graph of this function is shown in Fig. 55. As we can see, the error decreases with increasing T very slowly. The error is greater as F becomes smaller. Thus, e.g., in order for the error to be less than 10%, it is necessary to take FT = 100; with a resolving power of 100 cps the measurement should be prolonged no more than one second.

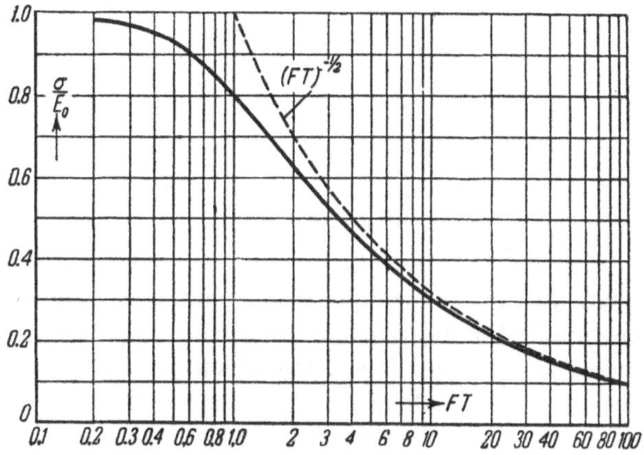

Fig. 55.

We note that Eq. (31.6) expresses the error in terms of the true spectrum $G(\omega)$, which is unknown, since it is just what we are to determine as a result of our measurements. The principal error that arises here is overcome by assuming that the spectral density is constant in the narrow band Ω. In principle it is impossible to compute the error only on the basis of the results of spectral measurement. In order to compute the error there is nothing more that need be known, other than the spectrum correlation, which is the moment of second order; it is therefore necessary to measure the moments of fourth order. The details can be found in special papers treating this problem [20, 21].

Let us now consider the problems of the procedure and technique of the spectral analysis of random processes. First of all we note that from the basic definition of the spectrum of a random process

$$G(\omega) = \frac{2}{\pi} \int_0^\infty B(\tau) \cos \omega\tau \, d\tau$$

we find that an indirect measurement of the spectrum may be accomplished by finding the correlation function $B(\tau)$ from experiment, and subsequently obtaining the spectrum by a Fourier transformation (which can be realized either by numerical processing of the experimental data or by means of some sort of apparatus that will automatically make the Fourier transformation). For this means of finding the spectrum the basic instrument applied is not an analyzer, but a correlation meter. There is an abundance of literature on the problems of measuring the correlation function, a survey of which would transcend the scope of the present work. We merely note that there is the existing opinion [6] that the determination of a spectrum through the correlation function is more economical from the point of view of the total number of measurements and computational operations required than a direct measurement of the spectrum.

As for the methods of direct analysis, the simplest analyzer is found in the form of an assembly of filters; integrating square-law instruments, which measure the energy, have to be used for the output indicators. This is the most uncluttered method of measurement, since the application of wattmeters essentially reduces the operation to a weighted integration; the weighting function depends on the properties of the instrument itself and the circuit to which it is hooked up, thus introducing an element of arbitrary judgement into the results of the measurement. More precisely, the result of the measurement is a quantity, the definition of which follows from the properties of the measuring apparatus. An example of such a specialized definition can be observed in the previously cited paper by Fano [22].

Another possible means for analysis is contained in the fact that the energy at the output of a low-frequency filter with a variable upper limit, i.e., the energy in a band ranging from 0 to ω, can be measured. This energy is equal to

$$E_T(\omega) = \int_0^\infty G_T(\nu)\, d(\nu).$$

If the function thus found is differentiated with respect to the frequency, we obtain

$$G_T(\omega) = \frac{\partial}{\partial\omega} E_T(\omega).$$

There is also a possible approach, which has for the most part been quite satisfactory in the technique of analysis, namely, a periodic repetition of a specimen of the analyzed process. This finds practical realization by means of a record, for example, on magnetic tape, with the subsequent splicing of a sufficiently long segment of the record into a loop. At the splice there will, of course, be a discontinuity, if not in the function itself, then in its derivatives. This fact shows up in the result of the analysis as a specific effect, which it is sometimes necessary to take into account [31].

32. On the Shrinkage of a Spectrum

Ever since the conception of the technology of electrical communications, one problem has never ceased to agitate specialists — the problem of reducing the width of the spectrum of a signal, in short, the problem of shrinking the spectrum. Numerous attempts have so far been unable to bring the problem to a really satisfactory solution as a whole. There are, needless to say, a number of papers representing definite advances in the special aspects of the problem. Without undertaking an analysis of these papers at this point, we shall restrict ourselves to a general presentation of the problem, in which so doing we shall be required to introduce the fundamental concepts of information theory.

It is quite obvious that the shrinking of the spectrum of some signal, considered as a random process $\xi(t)$, is possible only as the result of some transformation to which the original process is subjected. We shall form the new process

$$\eta(t)=\Psi[\xi(t)], \tag{32.1}$$

where Ψ is the symbol for some functional operation, and it is necessary to decide upon the operation Ψ in such a way that the process $\eta(t)$ will have a narrower spectrum than the original process $\xi(t)$. If the transformation is reversible, i.e., if there is a unique inverse operation Ψ^{-1} by means of which the original process can be reproduced, i.e.,

$$\xi(t)=\Psi^{-1}[\eta(t)]=\Psi^{-1}\{\Psi[\xi(t)]\}, \tag{32.2}$$

then, obviously, the problem is completely solved. It is necessary right from the start to point out that the requirement of reversibility is far from being fulfilled in all transformations. Thus, e.g., a spectrum-shrinking operation that is carried out by means of an ideal filter and that will completely eliminate a part of the spectrum of the process is irreversible.

The requirement of total reversibility, expressed in (32.2) is a very rigid one, and can be intelligently moderated by relying on the information representative of the original process. Specifically, it can be required that the operation Ψ be an invariant with respect to the information contained in the original process. In other words, in making the transformation, information should not be lost (for obvious reasons, it cannot increase, since in the transformation there is no information source involved). When the problem is presented in this way it is possible to think of the original process as depending on two functions, one of which contains all the information, the second of which contains no information at all. A function that is not an information bearer can be added to the reception end of a communication system; there is no need for this function to be transmitted. Thus the requirement for the operation Ψ is now reduced to the following two: 1) the transformation of the process η (t) should have a narrower spectrum and 2) the transformation of the process should preserve the information of the original process.

Let us now recall certain definitions that are applicable in information theory. We shall be concerned with quantum processes, i.e., processes with a discrete probability distribution, as the original process, since in this case the definitions are simple and unambiguous.

Let the original process be a random function with a bounded spectrum extending from 0 to F.[1] According to Kotel'nikov's theorem, such a process can be replaced by a sequence of instantaneous values read every interval of duration

$$\Delta t = \frac{1}{2F}.$$

Over the interval T these readings will be

$$n = \frac{T}{\Delta t} = 2FT.$$

Furthermore, let the instantaneous value be capable of assuming one of m equally probable values. There will be a total of

$$N = m^n \tag{32.3}$$

possible realizations. In this case the information contained in the process ξ (t) is defined as

$$I = \log_2 N = n \log_2 m, \tag{32.4}$$

or

$$I = 2FT \log m. \tag{32.5}$$

From this equation it is immediately evident that there are only two possibilities for shrinking the spectrum F with no change in the information I: either as the result of an increase in T or an increase in m. Such a transformation can be made in different ways; two simple examples are given below.

[1]Here we shall omit the fine points in connection with the informational aspect of Kotel'nikov's theorem (see the footnote on p. 81).

Example 1. Let the signal ξ (t) be recorded, after which it is reproduced with the speed of the carrier reduced. Here the spectrum is shrunk in the same proportion as the duration is increased. The original signal is reinstated by means of a repeated recording followed by reproduction with an increased speed. The necessary transformation can be written in the following form:

$$\eta(t) = \Psi[\xi(t)] = \xi(at),$$

$$\xi(t) = \Psi^{-1}[\eta(t)] = \eta\left(\frac{1}{a}t\right).$$

Example 2. A group of M figures of the original sequence can be represented by an M-digit number, written in a reference system with base m. But the same group can be written by a single-digit number in a system with base l > m, where, obviously, we must have

$$l \geqslant m^M.$$

With such a transformation during the same time it is necessary to transfer a single symbol instead of M symbols. This means that the width of the spectrum is shrunk by a factor of

$$M = \frac{\log l}{\log m}.$$

The possibilities for shrinking the width of a spectrum within the framework of Eq. (32.5), as we can see, is very limited. Transformations that involve the statistics of a signal are considerably more interesting.

The expression for the information contained in a sequence of n symbols can in general be written in the form

$$I = nI',$$

where I' is the information in one symbol (entropy). In the case already considered of m equally probable values

$$I' = \log m = I'_0. \tag{32.6}$$

If the values that the process can assume are not equally probable and the probability of the ith value is denoted by p_i, then the information in the symbol will be determined by

$$I' = -\sum_{i=1}^{m} p_i \log p_i = I'_1. \tag{32.7}$$

It is easily seen that when

$$p_i = \frac{1}{m}$$

Equation (32.7) goes over to (32.6). Equation (32.7) assumes that the symbols are not equally probable, but statistically independent. In the most general case, when there are statistical connections between the symbols, it is possible to characterize the statistical structure of the signal by a multidimensional probability of a given realization

$$p_k = p_k(\xi_1, \xi_2, \ldots, \xi_n)$$

and in this case

$$I' = -\frac{1}{n} \sum_{k=1}^{N=m^n} p_k \log p_k = I'_2. \quad (32.8)$$

Everything from this point on is based on the fact that

$$I'_0 > I'_1 > I'_2.$$

Therefore, if we write the expression for the information in general form

$$I = nI' = 2FTI',$$

then it becomes clear that by not varying the duration T and the number of levels m, it is possible to abbreviate the width of the spectrum F as the result of a statistical transformation of the statistics of the process, i.e., a change in the quantity I'. This transformation must result in the breaking of the statistical ties (transition from I'_2 to I'_1) with the subsequent equalization of the probabilities of the independent symbols (transition from I'_1 to I'_0). The information contained in a sequence of equally probable and independent symbols, i.e.,

$$I_0 = nI'_0 = 2FTI'_0$$

is the greatest amount; in other words, shrinkage of the spectrum on a statistical basis when $I' = I'_0$ is impossible.

The quantity

$$R = 1 - \frac{I'}{I'_0} \quad (32.9)$$

is called the residue. Thus, the above can be formulated as follows: statistical shrinkage of a spectrum is possible as a result of the elimination of the residue. For a fixed number of levels the expression for the residue can be written also in the form

$$R = 1 - \frac{F_0 T_0}{FT}, \tag{32.10}$$

where F and T are the actual width of the spectrum and duration of the signal, respectively, while F_0 and T_0 are the least values of the same quantities for an ideally transformed signal (i.e., for a signal for which $I' = I_0'$). From (32.10) it follows that with a fixed signal duration the width of the spectrum can be reduced by a factor of

$$\frac{F}{F_0} = \frac{1}{1-R}.$$

Thus, for the clarification of the possibility for the statistical shrinkage of a spectrum it is necessary that we be able to evaluate the magnitude of the residue, while for the realization of this possibility it is necessary to find a transformation that will eliminate, or, at least, cut down the residue. This expansive problem, from the point of view of information theory, amounts to the same thing as the problem of statistically optimum coding. The discussion of this means and the approach to its solution would lead us too far afield. We shall concern ourselves merely with one remark, which explains the principle involved.

We shall make use of the data associated with modulated oscillations (Section 30). It is known that the width of the spectrum of an amplitude-modulated oscillation is just twice as great as the width of the spectrum of the modulating process. Since as a result of modulation the amount of information does not vary, the modulated oscillation has the residue

$$R = 1 - \frac{F}{2F} = 0.5$$

(it is assumed that the modulating process has no residue). By means of detection the original (modulating) process is reestablished. Thus we can consider the modulation as a transformation which increases the residue (and which broadens the spectrum), while the detection is treated as an operation which eliminates the residue (and shrinks the spectrum). Consequently, if the statistics of the original process are similar to the statistics of an amplitude-modulated oscillation, then an operation of the detection type will give the desired shrunken spectrum. Another possibility for the elimination of the residue and for cutting the spectral width in half as well consists in eliminating one of the side bands of the modulation spectrum.

The same considerations are also applicable to an oscillation that is modulated in phase (or in frequency). The problem is somewhat complicated in the given case by the fact that the construction of the side bands is not a repetition of the construction of the modulating spectrum, as in the case of AM. Therefore, the width of the spectrum of an FM-oscillation must be determined beforehand. Let us take the case of a large index and a normally distributed modulating process, when the spectrum of the side band is written in the form [Eq. (30.11)]

$$G(\omega) = \sqrt{\frac{2}{\pi}}\frac{1}{\Delta\omega} e^{-\frac{\omega^2}{2\Delta\omega^2}}.$$

In order to determine the width of the spectrum we shall make use of the moment criterion (Section 12), i.e., we shall assume

$$\Omega_0^2 = \frac{I}{A} - \frac{M^2}{A^2},$$

where

$$A = \int_0^\infty G\,d\omega, \quad M = \int_0^\infty \omega G\,d\omega, \quad I = \int_0^\infty \omega^2 G\,d\omega.$$

Carrying out the integration, we obtain

$$A = 1, \quad M = \sqrt{\frac{2}{\pi}}\Delta\omega, \quad I = \Delta\omega^2,$$

whence

$$\Omega_0 = \sqrt{1 - \frac{2}{\pi}}\Delta\omega \simeq 0.8\,\Delta\omega,$$

and the total width of the spectrum is

$$2\Omega \simeq 1.6\,\Delta\omega \simeq 2\,\Delta\omega.$$

It can thus be considered that in order of magnitude the width of the spectrum of an FM-oscillation is determined by twice the frequency deviation (i.e., the swing band). If we assume

$$\Delta f = \frac{1}{2\pi} \Delta\omega = 75 \text{ kc}$$

then the width of the spectrum comes to

$$F \simeq 2\Delta f = 150 \text{ kc}$$

When the width of the spectrum of the modulating process is 5 kc

$$\frac{F}{F_0} = 30,$$

and the residue is

$$R = 1 - \frac{F}{F_0} \simeq 0.97.$$

Some researchers assume that speech is a process wherein the statistics are very near the statistics of an FM-oscillation with a large index. If this is so, then an operation of the frequency-detection type can lead to a considerable shrinkage of the speech spectrum. Work is presently being done in this direction.[1]

[1]See, e.g., the report by Marquis and Daré, Third Symposium on Information Theory (London, 1955).

APPENDIXES

I. On the Width of the Spectrum of a Product of Functions

In Section 4 we proved a theorem on the spectrum of a product of two functions of time: if two time functions $f_1(t)$ and $f_2(t)$ are given, along with their spectra $S_1(\omega)$ and $S_2(\omega)$, then the spectrum of the product of the functions $f_1(t)$ and $f_2(t)$ is expressed by the equation

$$G(\omega) = \frac{1}{2\pi} \int_{-\infty}^{+\infty} S_1(\nu) S_2(\omega - \nu) \, d\nu.$$

A similar relation exists also for the power spectra (see Section 27):

$$G(\omega) = \frac{1}{\pi} \int_{-\infty}^{\infty} G_1(\nu) G_2(\omega - \nu) \, d\nu. \tag{I.1}$$

As one of the applications of this equation we shall consider the problem of the width of the spectrum of a product of two functions.

The problem stands as follows: two functions are given, to which their spectra are bounded, i.e., they have a width $\Omega_1 = \omega_2 - \omega_1$ for the first function and $\Omega_2 = \omega_4 - \omega_3$ for the second. This property of the spectra specified by us can be written in the form

$$G_1(\omega) \neq 0 \text{ for } \omega_1 < \omega < \omega_2,$$
$$G_2(\omega) \neq 0 \text{ for } \omega_3 < \omega < \omega_4.$$

We must determine the width of the spectrum of the product of the functions. The spectrum of the product is not equal to zero, unless the integrand in Eq. (I.1) is equal to zero. This condition is satisfied when the spectra $G_1(\nu)$ and $G_2(\omega - \nu)$ are overlapping, i.e., in that region of values of ω for which over some interval of the values of ν, both of the spectra $G_1(\nu)$ and $G_2(\omega - \nu)$ are not equal to zero. From these considerations we can introduce a system of inequalities, giving us the answer to our question. The formation of this system is greatly facilitated by means of the graphs in Fig. 56.

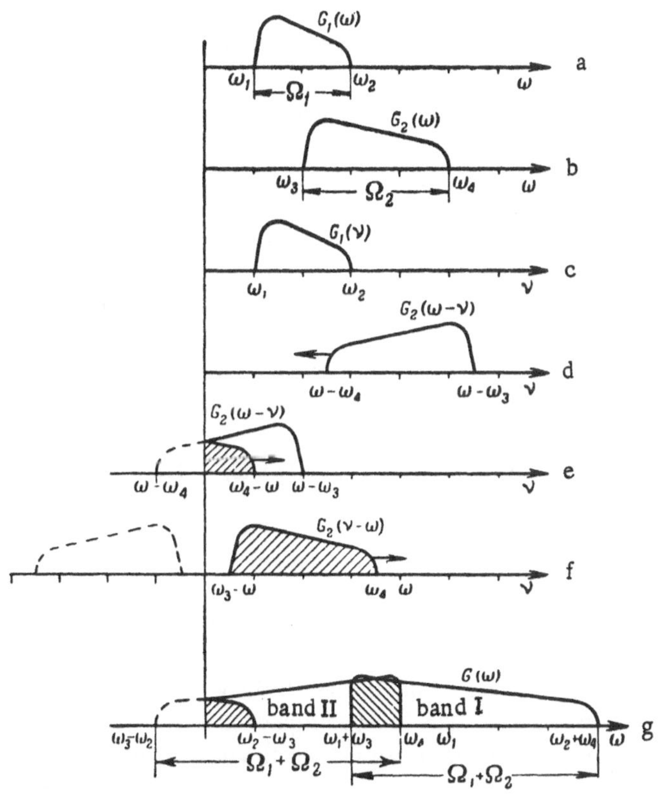

Fig. 56.

In Fig. 56a and b the given spectra $G_1(\omega)$ and $G_2(\omega)$ are shown. In Fig. 56c the spectrum $G_1(\nu)$ is given; this graph does not differ in any way from Fig. 56a. In Fig. 56d the spectrum $G_2(\omega - \nu)$ is constructed. It differs from Fig. 56b in that the reference origin is transfered to the point $\nu = \omega$ and the spectrum is turned around. ω is the running frequency of the resultant spectrum. When it is decreased, the spectrum moves to the left, as shown by the arrow. The condition for overlapping of the spectra in Fig. 56c and Fig. 56d can be written in the form

$$\omega_2 > \omega - \omega_4$$

or

$$\omega < \omega_2 + \omega_4. \tag{I.2}$$

The overlapping is continued as long as the following condition holds:

$$\omega - \omega_3 > \omega_1$$

or

$$\omega > \omega_1 + \omega_3. \tag{I.3}$$

Combining the inequalities (I.2) and (I.3), we can write for the first overlap band

$$\omega_1 + \omega_3 < \omega < \omega_2 + \omega_4, \tag{I.4}$$

from which the width of the first band is

$$\omega_2 + \omega_4 - (\omega_1 + \omega_3) = \Omega_1 + \Omega_2.$$

With a further decrease in ω the spectrum $G_2(\omega - \nu)$ enters the region of negative frequencies, as shown by the dashed curve in Fig. 56e. The actual spectrum though is situated in the region of positive frequencies; it can be constructed by bending the spectrum around the ordinate axis. The twisted part of the spectrum is striated in Fig. 56e. With a further decrease in the frequency ω the inverted spectrum moves all the way along the ν-axis to the right, as shown in Fig. 56f. The last two figures show that the spectra enter a second overlap band, the boundaries of which are determined by the inequalities

$$\omega_1 < \omega_4 - \omega, \quad \omega_2 > \omega_3 - \omega,$$

i.e.,

$$\omega_3 - \omega_2 < \omega < \omega_4 - \omega_1. \tag{I.5}$$

The width of the second band is

$$\omega_4 - \omega_1 - (\omega_3 - \omega_2) = \Omega_1 + \Omega_2.$$

Thus we have obtained the following rule:

The spectrum of a product of two functions consists of two bands, the width of each being equal to the sum of the widths of the spectra of both constituents. The boundaries of the bands are determined by the inequalities (I.4) and (I.5).

This still does not mean that the over-all width of the spectrum of the product is simply twice the sum of the widths of the spectra of the constituents. The fact is that one of the bands can take in a part of the negative-frequency region; in this case it is bent around, as described above, and its actual width is foreshortened. Moreover, both bands can be partially overlapping.

The whole picture is shown in Fig. 56g. All the graphs of Fig. 56 are assumed to have the following numerical values:

$$\omega_1 = 1,\ \omega_2 = 3,\ \omega_3 = 2,\ \omega_4 = 5,\ \Omega_1 = 2,\ \Omega_2 = 3,$$
$$\omega_2 + \omega_4 = 8,\ \omega_1 + \omega_3 = 3,\ \omega_4 - \omega_1 = 4,\ \omega_3 - \omega_2 = -1.$$

The spectrum of the product in this example occupies the unbroken band

$$0 < \omega < 8.$$

Therefore, the width of the spectrum of the product can be determined by means of the following rule:

The over-all width of the spectrum of a product of two functions is never less than the sum, but never greater than twice the sum of the widths of the spectra of both constituents.

The rules given above have a quite general character. They can, for example, be applied to discrete (line) spectra as well. Two additional rules can also be introduced, with reference to discrete spectra alone:

The number of lines in each of the two bands of the spectrum of a product of two periodic functions is equal to the product of the number of lines in the spectra of the constituents.

However, the individual lines can merge. Therefore:

The total number of lines in the spectrum of a product of two periodic functions is never less than the product, but never greater than twice the product of the number of lines in the spectra of the constituents.

For the application of these rules it should be borne in mind that the presence of a constant component is expressed as a spectral line with frequency $\omega = 0$.

As an example we shall consider the spectrum of a mixed-modulation oscillation. Let the carrier frequency ω_0 be frequency-modulated with a frequency ω_1, then amplitude-modulated with a frequency ω_2. Analytically such an oscillation is expressed by the equation

$$x = c_0 (1 + m \sin \omega_2 t) \sin (\omega_0 t + \beta \sin \omega_1 t),$$

where c_0 is the carrier amplitude, $\underline{m}$ is the AM depth, β is the FM index.

If we write the notation

$$\sin (\omega_0 t + \beta \sin \omega_1 t) = f_1 (t),$$
$$1 + m \sin \omega_2 t = f_2 (t),$$

then the problem of the spectrum of the mixed-modulation oscillation reduces to the problem of the spectrum of a product. The spectrum $G_1(\omega)$ of the function $f_1(t)$ when the index is small consists of three lines with frequencies ω_0, $\omega_0 \pm \omega_1$. The spectrum $G_2(\omega)$ of the function $f_2(t)$ consists of two lines with frequencies 0 and ω_2. In order to obtain the resultant spectrum it is necessary to move the spectrum $G_2(\omega - \nu)$, which has two lines with frequencies ω and $\omega - \omega_2$, along the ν-axis. In the resultant spectrum lines will appear every time that any two lines of the spectra $G_1(\nu)$ and $G_2(\omega - \nu)$ coincide. Noting all the possible coincidences, we find that the spectrum of the mixed-modulation oscillation considered will consist of ten lines with the frequencies

$$\omega_0,\ \omega_0 \pm \omega_1,\ \omega_0 \pm \omega_2,\ \omega_0 \pm \omega_1 \pm \omega_2.$$

The extreme frequencies of the spectrum are $\omega_0 \pm (\omega_1 + \omega_2)$.

II. The Spectra of Certain Frequency-Modulated Oscillations

In Section 7 we calculated the spectrum for a simple case of sinusoidal FM:

$$\omega = \omega_0 \left(1 + \frac{\Delta\omega}{\omega_0} \cos \Omega t\right).$$

The general case of FM can be written in the form

$$\omega = \omega_0 [1 + m f (t)],$$

whence

$$\theta = \omega_0 t + m F (t),$$

where

$$F(t)=\int_0^t f(\tau)\,d\tau.$$

Since the modulated oscillation has the form

$$x=\sin\vartheta=\sin[\omega_0 t+mF(t)],$$

It becomes obvious that the representation of the function F(t) by any arbitrary series (for example, by a power or trigonometric series) will lead to nothing satisfactory. It is necessary to seek a representation for F(t) in finite form, and, moreover, in such a way that the integral used to express the spectrum can be computed. We shall give two examples of such calculations here.

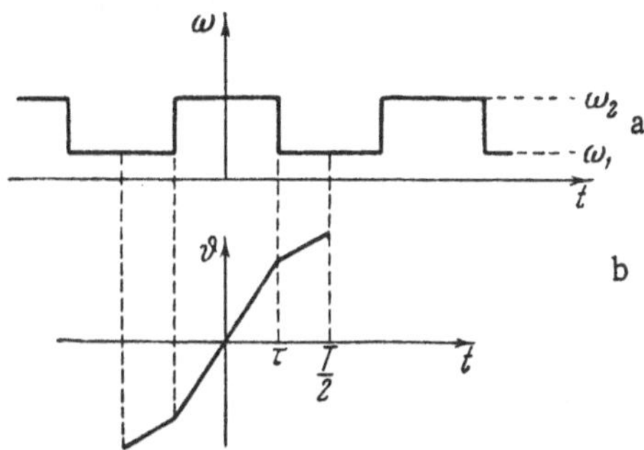

Fig. 57.

a. Spectrum of an oscillation with the frequency modulated by a square wave. The frequency varies according to the following (Fig. 57a):

$$\omega=\omega_2 \text{ when } -\tau<t<\tau,$$
$$\omega=\omega_1 \text{ when } -\frac{T}{2}<t<-\tau,\ \tau<t<\frac{T}{2}$$

etc. In Fig. 57b the graph for ϑ is given, where the analytic expression for the latter on the interval $\left(-\frac{T}{2},\frac{T}{2}\right)$ is the following:

$$\vartheta = \omega_2 t \quad \text{when} \quad -\tau < t < \tau,$$
$$\vartheta = \omega_1 t + b\tau \quad \text{when} \quad \tau < t < \frac{T}{2},$$
$$\vartheta = \omega_1 t - b\tau \quad \text{when} \quad -\frac{T}{2} < t < -\tau,$$

where $b = \omega_2 - \omega_1$. Since ϑ is an odd function, the spectrum of the function $x = \sin\vartheta$ is expressed in terms of the sine coefficients:

$$c_k = b_k = \frac{1}{T}\left[\int_{-\tau}^{\tau} \sin\omega_2 t \sin k\Omega t\, dt + \int_{-\frac{T}{2}}^{-\tau} \sin(\omega_1 t - b\tau)\sin k\Omega t\, dt + \right.$$
$$\left. + \int_{\tau}^{\frac{T}{2}} \sin(\omega t_1 + b\tau)\sin k\Omega t\, dt\right].$$

Performing some simple transformations and integrating, we find

$$c_k = \frac{1}{T}\left\{\frac{1}{\omega_2 - k\Omega}\sin(\omega_2 - k\Omega)\tau - \frac{1}{\omega_2 + k\Omega}\sin(\omega_2 + k\Omega)\tau + \right.$$
$$+ \frac{1}{\omega_1 - k\Omega}\left[\sin\left(\omega_1\frac{T}{2} + b\tau - k\pi\right) - \sin(\omega_2 - k\Omega)\tau\right] -$$
$$\left. - \frac{1}{\omega_1 + k\Omega}\left[\sin\left(\omega_1\frac{T}{2} + b\tau + k\pi\right) - \sin(\omega_2 + k\Omega)\tau\right]\right\}.$$

In the special case $\tau = \frac{T}{4}$, i.e., when the up and down teeth in the square wave (Fig. 57a) have the same width, we obtain

$$c_k = \frac{1}{4}\left\{\frac{\sin\frac{\pi}{2}\left(\frac{\omega_2}{\Omega} - k\right)}{\frac{\pi}{2}\left(\frac{\omega_2}{\Omega} - k\right)} - \frac{\sin\frac{\pi}{2}\left(\frac{\omega^2}{\Omega} + k\right)}{\frac{\pi}{2}\left(\frac{\omega_2}{\Omega} + k\right)} + \right.$$
$$+ \frac{\sin\frac{\pi}{4}\left(\frac{\omega_1}{\Omega} - k\right)}{\frac{\pi}{4}\left(\frac{\omega_1}{\Omega} - k\right)}\cos\frac{\pi}{4}\left(\frac{\omega_1}{\Omega} + 2\frac{\omega_2}{\Omega} - 3k\right) +$$
$$\left. + \frac{\sin\frac{\pi}{4}\left(\frac{\omega_1}{\Omega} + k\right)}{\frac{\pi}{4}\left(\frac{\omega_1}{\Omega} + k\right)}\cos\frac{\pi}{4}\left(\frac{\omega_1}{\Omega} + 2\frac{\omega_2}{\Omega} + 3k\right)\right\}.$$

In Fig. 58 the spectrum computed from this equation is shown. This spectrum has maxima at $k\Omega = \omega_1$ and $k\Omega = \omega_2$, as should be expected.

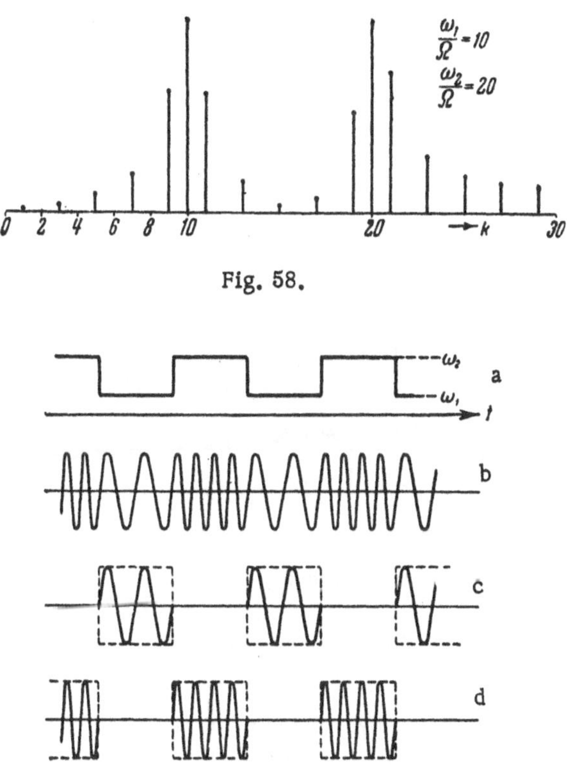

Fig. 58.

Fig. 59.

The spectrum of the FM-oscillation that we are considering can be computed on the basis of other considerations. In Fig. 59a the same frequency-variation law as in Fig. 57a is shown, while in Fig. 59b the corresponding FM-oscillation is shown. This oscillation can be expanded into the two oscillations shown in Fig. 59c and 59d. But each of these components turns out to be nothing more than the oscillation of the corresponding carrier frequency, modulated in amplitude by rectangular pulses. The spectrum of such an oscillation is found without any difficulty, after which the spectrum of the FM-oscillation is found as the sum of the spectra of both AM-oscillations. This example can be applied in that case when the frequency assumes, rather than just two, any number of fixed oscillations. Consequently, it is possible in this way to compute the spectra, not only for the signals of the ordinary frequency-type telegraph, but also for the two-channel frequency-type telegraph, which operates at four fixed frequencies. The computational technique does not change when there

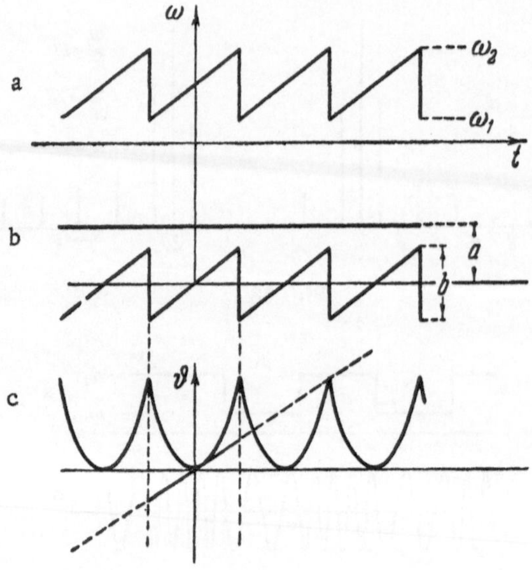

Fig. 60.

is a phase jump upon transition from frequency to frequency; in this case the calculations are, of course, made somewhat more complicated.

b. Spectrum of a frequency-modulated oscillation with the frequency varying according to a sawtooth law. This case — of the so-called "sweep signal" — is a little more complicated than the preceding one.

Let

$$\omega = a + \frac{b}{T} t \quad \text{for} \quad -\frac{T}{2} < t < \frac{T}{2}$$

(Fig. 60a), where $a = \frac{\omega_1 + \omega_2}{2}$, $b = \omega_2 - \omega_1$. In Fig. 60b both components of the frequency are shown separately.

For the argument ϑ we obtain

$$\vartheta = at + \frac{b}{2T} t^2 \quad \text{for} \quad -\frac{T}{2} < t < \frac{T}{2}$$

(Fig. 60c), and the modulated oscillation is written in the form

$$x = \sin\left(at + \frac{b}{2T}t^2\right).$$

Let us compute the components of the spectrum:

$$a_k = \frac{2}{T}\int_{-\frac{T}{2}}^{\frac{T}{2}} \sin\left(at + \frac{b}{2T}t^2\right)\cos k\Omega t\,dt,$$

$$b_k = \frac{2}{T}\int_{-\frac{T}{2}}^{\frac{T}{2}} \sin\left(at + \frac{b}{2T}t^2\right)\sin k\Omega t\,dt.$$

In order to compute these integrals it is first necessary to make some trigonometric transformations.

For example, for a_k we obtain

$$a_k = \frac{1}{T}\int_{-\frac{T}{2}}^{\frac{T}{2}} [\sin(\alpha t + \gamma^2 t^2) + \sin(\beta t + \gamma^2 t^2)]\,dt,$$

where for the sake of brevity the following notation is introduced:

$$\alpha = a + k\Omega, \quad \beta = a - k\Omega, \quad \gamma^2 = \frac{b}{2T}.$$

In this form it is necessary to add terms to the argument in order to complete the square, adding and subtracting $\frac{\alpha^2}{4\gamma^2}$ and $\frac{\beta^2}{4\gamma^2}$, respectively. Then, once again expanding the sines of the difference, we find

$$a_k = \frac{1}{2T\gamma}\left(\cos\frac{\beta^2}{4\gamma^2}C_y + \sin\frac{\beta^2}{4\gamma^2}S_y - \cos\frac{\alpha^2}{4\gamma^2}C_x - \sin\frac{\alpha^2}{4\gamma^2}S_x\right).$$

Here the following notation has been used:

$$C_x = \int_{x_1}^{x_2} \cos x^2\, dx, \qquad C_y = \int_{y_1}^{y_2} \cos y^2\, dy,$$

$$S_x = \int_{x_1}^{x_2} \sin x^2\, dx, \qquad S_y = \int_{y_1}^{y_2} \sin y^2\, dy,$$

$$x_1 = \frac{1}{2}\left(\frac{\alpha}{\gamma} - \gamma T\right), \qquad y_1 = \frac{1}{2}\left(\frac{\beta}{\gamma} - \gamma T\right),$$

$$x_2 = \frac{1}{2}\left(\frac{\alpha}{\gamma} + \gamma T\right), \qquad y_2 = \frac{1}{2}\left(\frac{\beta}{\gamma} + \gamma T\right).$$

A similar equation is obtained for b_k:

$$b_k = \frac{1}{T\gamma}\left(\cos\frac{\beta^2}{4\gamma^2} S_y - \sin\frac{\beta_2}{4\gamma^2} C_y + \cos\frac{\alpha^2}{4\gamma^2} S_x - \sin\frac{\alpha^2}{4\gamma^2} C_x\right).$$

We are interested in the amplitude spectrum. The amplitude of the kth harmonic is equal to ·

$$c_k = \sqrt{a_k^2 + b_k^2}\,.$$

Performing the computations, we obtain

$$c_k = \frac{1}{T\gamma}\left[C_y^2 + S_y^2 + C_x^2 + S_x^2 - 2\cos\frac{\beta^2 + \alpha^2}{4\gamma^2}(C_y C_x - S_y S_x) - \right.$$
$$\left. - 2\sin\frac{\beta^2 + \alpha^2}{4\gamma^2}(C_y S_x + S_y C_x)\right]^{\frac{1}{2}}.$$

In conclusion, we note that the spectrum of an FM-oscillation (as opposed to AM) in general is not symmetric about a center frequency, had the latter even been defined; the character of the asymmetry in the spectrum depends on the law of variation of the frequency. A symmetric spectrum is obtained only when the frequency variation is symmetric (about some central value). The sinusoidal variation in frequency considered in Section 7 can serve as an example.

III. Active Band of a Spectrum

Certain ideas in connection with the instantaneous spectrum evolve from the concept of the active band of a spectrum, as proposed by D. V. Ageev [1].

We shall proceed from the definition of the instantaneous power spectrum according to Page (see Section 6):

$$\rho(\omega, t) = \frac{\partial}{\partial t} | S_t(\omega) |^2,$$

where

$$S_t(\omega) = \int_{-\infty}^{t} f(u)\, e^{-j\omega u}\, du$$

is the running spectrum. By definition the instantaneous power is

$$f^2(t) = \frac{1}{\pi} \int_0^{\infty} \rho(\omega, t)\, d\omega.$$

However, it is possible to determine some frequency interval within whose limits the greater part of the power is concentrated. This condition can be written in the form

$$\frac{1}{\pi} \int_{\omega_1}^{\omega_2} \rho(\omega, t)\, d\omega = \eta f^2(t),$$

where η is a proper fraction, somewhat less than unity. The interval (ω_1, ω_2) can be called the active band. We made use of a similar criterion in Section 12, but now we are concerned with instantaneous spectra, so that, consequently, ω_1 and ω_2 are time functions. Therefore, not only the width of the interval (ω_1, ω_2), but its position on the frequency scale as well depends on the time.

It is possible to arrive at a definition of the active band in another way. From the definition of the running spectrum

$$f(t) = \frac{1}{2\pi} \int_{-\infty}^{\infty} S_t(\omega)\, e^{j\omega t}\, d\omega = \frac{1}{\pi} \int_0^{\infty} (A_t \cos \omega t - B_t \sin \omega t)\, d\omega,$$

where

$$A_t + jB_t = S_t(\omega).$$

Let us now introduce the function $f_1(t)$, which is expressed by a similar interval, but within finite limits:

$$f_1(t) = \frac{1}{\pi} \int_{\omega_1}^{\omega_2} (A_t \cos \omega t + B_t \sin \omega t)\, d\omega$$

and let us demand that the difference

$$\Delta f = f(t) - f_1(t)$$

be (from the point of view of some definite criterion) sufficiently small, i.e., that the function $f_1(t)$ serve as a satisfactory approximation to f(t). The interval (ω_1, ω_2) will in this case be defined as the active band of the spectrum.

In the paper cited [1] the following function is treated as an example:

$$f(t) = \sin \frac{1}{2} \lambda t^2,$$

i.e., an FM-oscillation whose frequency varies according to the linear law

$$\omega = \lambda t.$$

It is shown that the width of the active band is expressed by the relation

$$\omega_2 - \omega_1 = \Delta\omega = b \sqrt{\frac{d\omega}{dt}} = b\sqrt{\lambda},$$

where b is a coefficient that becomes larger as the approximation demanded becomes closer. It is also shown that the center frequency of the active band coincides with the instantaneous frequency, so that in the case considered the active band, while retaining a constant width, slides along the frequency scale, following the changing instantaneous frequency ω; this result falls in complete agreement with our intuitive ideas.

IV. Expansion of Spectra into the Spectra of Component Functions

The linearity of the Fourier transformation expressed by Eq. (4.1) permits the application of a general method of approximate computation of a spectrum; the method is based on the following considerations.

Let a given function f(t) be approximated by a finite sum of some arbitrarily chosen functions f_k, i.e.,

$$f(t) \approx \sum_{k=1}^{N} f_k(t).$$

If the spectra of the functions f_k are known, then, denoting them by S_k, we have

$$S = \int_{-\infty}^{+\infty} f(t)\, e^{-j\omega t}\, dt \approx \int_{-\infty}^{+\infty} e^{-j\omega t} \left(\sum_{k=1}^{N} f_k(t) \right) dt =$$

$$= \sum_{k=1}^{N} \int_{-\infty}^{+\infty} f_k e^{-j\omega t} dt = \sum_{k=1}^{N} S_k.$$

This relation can serve as the basis for innumerable variants of the equations and tables for computing the spectra of the given functions.

We shall limit our discussion here to two examples, which refer to non-periodic functions of a pulse character, in particular, to a function that is not equal to zero over some interval of time τ.

Let, for example, the function exist on the interval

$$-\frac{\tau}{2} < t < \frac{\tau}{2}.$$

Let us choose an expansion into trigonometric functions, i.e., let us expand f(t) in a Fourier series over the indicated interval. We then obtain

$$f(t) \approx \sum_{-N}^{N} C_k e^{j2\pi k \frac{t}{\tau}},$$

where

$$C_k = \frac{1}{\tau} \int_{-\frac{\tau}{2}}^{\frac{\tau}{2}} f(t)\, e^{-j2\pi k \frac{t}{\tau}}\, dt.$$

Thus in the present example the elementary function is

$$f_k = C_k e^{j2\pi k \frac{t}{\tau}}.$$

The spectrum of this function has the form

$$S_k = C_k \int_{-\frac{\tau}{2}}^{\frac{\tau}{2}} e^{-j\left(\omega - 2\pi\frac{k}{\tau}\right)t}\,dt = \frac{(-1)^k\,2C_k}{\omega - 2\pi\frac{k}{\tau}}\sin\omega\frac{\tau}{2},$$

and, consequently, the spectrum that we seek is

$$S \approx 2\sin\omega\frac{\tau}{2}\sum_{-N}^{N}\frac{(-1)^k C_k}{\omega - 2\pi\frac{k}{\tau}}.$$

The approximation will plainly be better as the number of components in the sum becomes greater.

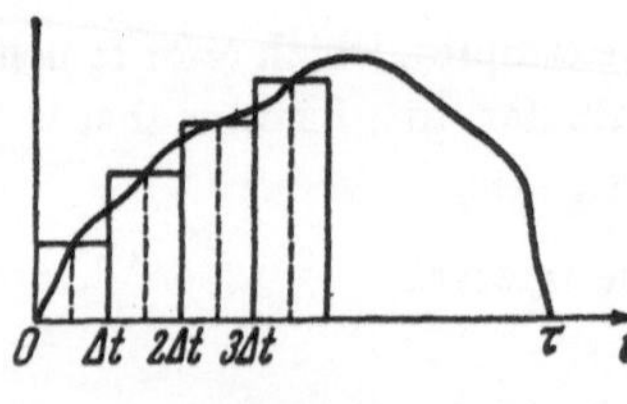

Fig. 61.

For the second example let us approximate the given function by a step curve, which divides the interval $0 < t < \tau$ into $\underline{n}$ parts of length Δt, i.e., assuming

$$\tau = n\,\Delta t,$$

and taking the elementary function in the form of a rectangular pulse. Then we can write (Fig. 61)

$$f(t) \approx \sum_{k=0}^{n-1}\{\sigma_0[t - k\,\Delta t] - \sigma_0[t - (k+1)\,\Delta t]\}\,f\left[\left(k + \frac{1}{2}\right)\Delta t\right].$$

The elementary pulse is equal to

$$f_k(t) = f\left[\left(k + \frac{1}{2}\right)\Delta t\right]\{\sigma_0[t - k\,\Delta t] - \sigma_0[t - (k+1)\,\Delta t]\}.$$

Its spectrum has the form

$$S_k = f\left[\left(k + \frac{1}{2}\right)\Delta t\right]\int_{k\,\Delta t}^{(k+1)\,\Delta t} e^{-j\omega t}\,dt = \frac{2}{\omega}e^{-j\omega\left(k+\frac{1}{2}\right)\Delta t}\sin\omega\frac{\Delta t}{2};$$

consequently, the desired spectrum has the form

$$S = \Delta t \frac{\sin \omega \frac{\Delta t}{2}}{\omega \frac{\Delta t}{2}} \sum_{k=0}^{n-1} f\left[\left(k + \frac{1}{2}\right)\Delta t\right] e^{-j\omega\left(k+\frac{1}{2}\right)\Delta t}.$$

The approximation becomes better with increasing n. In the limit, as $n \to \infty$, i.e., as $\Delta t \to 0$, the factor $\dfrac{\sin \omega \frac{\Delta t}{2}}{\omega \frac{\Delta t}{2}}$ goes to unity, and the sum goes over to the Fourier integral, which gives the exact value of the spectrum of the function f(t), i.e.,

$$S = \int_0^{\tau} f(t) e^{-j\omega t}\, dt.$$

V. The Spectrum of a Short Pulse with Alternating Sign

The assumption expressed in Section 11 concerning the spectrum of a short pulse, that the spectrum of such a pulse was uniform up to a frequency for which the period was comparable with the pulse duration, requires some refinement. The fact is that this assumption is valid when for the duration of the pulse the function expressing it does not change sign. i.e., for example, when on the interval $\left(-\frac{\tau}{2}, \frac{\tau}{2}\right)$ the pulse is defined by an even function. If, on the other hand, this condition is not satisfied, i.e., if, for example, over the indicated interval the pulse is defined by an odd function, then the assumption changes radically.

For the spectrum of a pulse we have

$$S = \int_{-\frac{\tau}{2}}^{\frac{\tau}{2}} f(t) e^{-j\omega t}\, dt.$$

Let us divide the integral into two integrals, and let the function f(t) be odd. Then

$$S=\int_{-\frac{\tau}{2}}^{0} f(t)e^{-j\omega t}\,dt+\int_{0}^{\frac{\tau}{2}} f(t)e^{-j\omega t}\,dt=\int_{0}^{\frac{\tau}{2}} f(t)(e^{j\omega t}-e^{-j\omega t})\,dt=$$

$$=2j\int_{0}^{\frac{\tau}{2}} f(t)\sin\omega t\,dt\approx 2j\omega\int_{0}^{\frac{\tau}{2}} tf(t)\,dt=j\omega M,$$

where M is the moment of first order of the function f(t) relative to the origin. The approximation is legitimate when $\omega\frac{\tau}{2}\ll 1$. Thus, up to frequencies determined by the comparability of the period and duration, the spectrum of a sign-changing pulse is expressed by a linear function of the frequency.

As an example let us compare the spectra of segments of a cosine and sine wave. For the cosinusoidal pulse we have

$$S=2\int_{0}^{\frac{\tau}{2}}\cos\Omega t\cos\omega t\,dt=\frac{\sin(\Omega+\omega)\frac{\tau}{2}}{\Omega+\omega}+\frac{\sin(\Omega-\omega)\frac{\tau}{2}}{\Omega-\omega}\approx\tau.$$

For the sinusoidal pulse we have

$$S=-2j\int_{0}^{\frac{\tau}{2}}\sin\Omega t\sin\omega t\,dt=2j\left(\frac{\sin(\Omega+\omega)\frac{\tau}{2}}{\Omega+\omega}-\frac{\sin(\Omega-\omega)\frac{\tau}{2}}{\Omega-\omega}\right).$$

Here we must make use of at least two terms of the expansion of the sine in a power series, in order to pass to the limit with decreasing τ. This gives

$$2j\left(\frac{\sin(\Omega+\omega)\frac{\tau}{2}}{\Omega+\omega}-\frac{\sin(\Omega-\omega)\frac{\tau}{2}}{\Omega-\omega}\right)\approx j\frac{\tau^{3}}{4\cdot 3!}[(\Omega+\omega)^{2}-(\Omega-\omega)^{2}]=j\omega\frac{\tau^{3}}{3!}\Omega.$$

In order to clarify the relations obtained we can further agree that every odd function can be treated as the derivative of some even function. However, the spectrum of the derivative is obtained by multiplication of the spectrum of the function by $j\omega$ [see (4.2)]. Consequently, if the spectrum of a short even pulse is uniform, i.e., is expressed by a constant quantity, then the spectrum of an odd pulse must contain the factor $j\omega$.

Finally, the spectra of discontinuous functions should also be indicated (Section 11). The spectrum of the unit pulse $\delta(t)$, which is an even function, is equal to unity. The spectrum of the second-order pulse $\frac{d}{dt}\delta(t)$, which is an odd function and which is characterized by two successive jumps to $+\infty$ and $-\infty$ about $t = 0$, is equal to $j\omega$.

If a short pulse makes a change of sign, then its spectrum can be expressed by higher powers of the frequency as well.

VI. Details of the Calculation of Δf and Δt

For the reader who wishes to carry through the computations of Section 12 or perform similar ones, some of the details of these computations are given below.

a. Rectangular pulse. $\Delta t = 0.9\tau$ – this is obvious. Furthermore,

$$\Phi = \tau \frac{\sin \omega \frac{\tau}{2}}{\omega \frac{\tau}{2}}; \quad \int_0^{\Delta\omega} \Phi^2 d\omega = \tau^2 \int_0^{\Delta\omega} \frac{\sin^2 \omega \frac{\tau}{2}}{\left(\omega \frac{\tau}{2}\right)^2} d\omega = 2\tau \int_0^{\Delta\omega \frac{\tau}{2}} \frac{\sin^2 z}{z^2} dz.$$

Here and in what follows are integrals of the form

$$\int \frac{\begin{matrix}\sin\\ \cos\end{matrix} z}{z^n} dz,$$

which are calculated by parts until they are terminated on the component $\mathrm{Si}(z)$ or $\mathrm{Ci}(z)$. As a result, for $\Delta\omega\tau$ we obtain the equation

$$\frac{2}{\pi}\left[\mathrm{Si}(\Delta\omega\tau) - \frac{\sin^2 \Delta\omega \frac{\tau}{2}}{\Delta\omega \frac{\tau}{2}}\right] = \eta = 0.9,$$

the solution of which is easily obtained by the method of pseudo-roots and turns out to be

$$\Delta\omega\tau = 5.1.$$

b. Triangular pulse. For a triangular pulse, starting from the condition

$$2\int_0^{\frac{\Delta t}{2}} \left(1-2\frac{t}{\tau}\right)^2 dt = \eta\frac{\tau}{3}$$

we obtain the equation

$$\left(\frac{\Delta t}{\tau}\right)^3 + 3\left(\frac{\Delta t}{\tau}\right)^2 - 3\frac{\Delta t}{\tau} = \eta = 0.9,$$

which in solving gives

$$\frac{\Delta t}{\tau} = 0.541.$$

The spectrum of a triangular pulse is

$$\Phi = \tau\frac{1-\cos\omega\frac{\tau}{2}}{\left(\omega\frac{\tau}{2}\right)^2}.$$

Writing the expression

$$\tau^2\int_0^{\Delta\omega} \frac{\left(1-\cos\omega\frac{\tau}{2}\right)^2}{\left(\omega\frac{\tau}{2}\right)^4}\, d\omega = \eta\frac{2}{\pi}\tau$$

and performing the integration, we obtain the equation

$$\frac{4}{\pi}\left\{\operatorname{Si}(x) - \frac{1}{2}\operatorname{Si}\left(\frac{x}{2}\right) - \frac{2}{x^3}\left[3 + \left(\cos x - 4\cos\frac{x}{2}\right) - \right.\right.$$
$$\left.\left. - \frac{x}{2}\left(\sin x - 2\sin\frac{x}{2}\right) - \frac{x^2}{2}\left(\cos x - \cos\frac{x}{2}\right)\right]\right\} = 0.9,$$

the solution of which is

$$x = \Delta\omega\tau = 5.3.$$

c. Cosine pulse. In order to determine Δt we have the relation

$$2\int_0^{\frac{\Delta t}{2}} \cos^2 \pi \frac{t}{\tau}\, dt = \eta \frac{\tau}{2},$$

which leads to the equation

$$\frac{\Delta t}{\tau} + \frac{1}{\pi} \sin \pi \frac{\Delta t}{\tau} = 0.9.$$

From this we find

$$\frac{\Delta t}{\tau} = 0.596\,.$$

The spectrum of a cosine pulse is

$$\Phi = \frac{\pi}{2}\tau \frac{\cos \omega \frac{\tau}{2}}{\left(\frac{\pi}{2}\right)^2 - \left(\omega\frac{\tau}{2}\right)^2},$$

and the expression for determining Δω assumes the form

$$\frac{\pi^2}{4}\tau^2 \int_0^{\Delta\omega} \frac{\cos^2 \omega \frac{\tau}{2}}{\left[\left(\frac{\pi}{2}\right)^2 - \left(\omega\frac{\tau}{2}\right)^2\right]^2} d\omega = \eta \frac{\pi}{2}\tau\,.$$

The integral appearing in this equation is computed by means of an expansion of the integrand into simple fractions. As a result of the integration we obtain the expression

$$\frac{1}{\pi}\left[-\frac{1-\cos x}{x} + \mathrm{Si}\,(x) + \frac{1}{\pi}\ln x - \frac{1}{\pi}\mathrm{Ci}\,(x)\right]\Bigg|_{x_1}^{x_2} = 0.9,$$

where

$$x_1 = \pi - \Delta\omega\tau, \quad x_2 = \pi + \Delta\omega\tau.$$

One difficulty arises here: x_1 is undoubtedly negative, and we cannot substitute it as an argument of either the logarithm or the symbol Ci. This difficulty, however, is very easily handled. Let us consider the function

$$\varphi(x) = \ln x - \mathrm{Ci}\,(x).$$

Recalling the well known expansion for Ci(x) in a power series, we obtain

$$\varphi(x) = -E + \frac{x^2}{2\cdot 2!} - \frac{x^4}{4\cdot 4!} + \frac{x^6}{6\cdot 6!} - \ldots,$$

where E is the Eulerian constant. Thus it turns out that $\varphi(x)$ is an even function and, consequently, we have the right to insert positive arguments in place of the negative ones after the symbols ln and Ci in carrying out the calculations. Bearing in mind, moreover, that Si(x) is an odd function, and performing all the calculations we find $\Delta\omega\tau = 4.57$.

VII. On a General Criterion for Determining Δf and Δt

The general criterion for evaluating Δf and Δt, which was used in Section 12 and which was based on the concept of the radius of inertia of a plane figure, proves to be very stringent and its application is not always possible.

On the basis of this criterion let us compute the product $\Delta f \Delta t$ for a triangular pulse, for which

$$f(t) = 1 - 2\frac{|t|}{\tau},$$

$$\Phi(\omega) = \frac{\tau}{2}\,\frac{1-\cos\omega\frac{\tau}{2}}{\frac{1}{2}\left(\omega\frac{\tau}{2}\right)^2}.$$

We find

$$A_t = 2\int_0^{\frac{\tau}{2}}\left(1-2\frac{t}{\tau}\right)^2 dt = \frac{13}{12}\tau, \quad I_t = 2\int_0^{\frac{\tau}{2}} t^2\left(1-2\frac{t}{\tau}\right)^2 dt = \frac{49}{480}\tau^3,$$

$$M_t = 0,$$

$$\Delta t^2 = \frac{I_t}{A_t} = 9.43\cdot 10^{-2}\tau^2,$$

$$\Delta t = 0.307\tau.$$

Let us now find the width of the spectrum:

$$A_\omega = \pi A_t = 3.4\tau,$$

$$M_\omega = \frac{\tau^3}{4}\int_0^\infty \omega\,\frac{\left(1-\cos\omega\frac{\tau}{2}\right)^2}{\frac{1}{4}\left(\omega\frac{\tau}{2}\right)^4}\,d\omega = 4\int_0^\infty \frac{\sin^4 x}{x^3}\,dx = \frac{8}{3},$$

$$I_{\omega}=\frac{\tau^2}{4}\int_0^{\infty}\omega^2\frac{\left(1-\cos\omega\frac{\tau}{2}\right)^2}{\frac{1}{4}\left(\omega\frac{\tau}{2}\right)^4}d\omega=\frac{16}{\tau}\int_0^{\infty}\frac{\sin^2 x}{x^2}dx=\frac{8\pi}{\tau}.$$

Thus

$$\Delta\omega^2=\frac{I_{\omega}}{A_{\omega}}-\left(\frac{M_{\omega}}{A_{\omega}}\right)^2=\frac{6.8}{\tau^2},\quad \Delta\omega=\frac{2.6}{\tau},\quad \Delta f=\frac{0.415}{\tau}$$

and, finally,

$$\Delta f\,\Delta t=0.415\cdot 0.307=0.127,$$

i.e., the result is approximately three times as large as the theoretical minimum.

If we now set out to perform similar calculations for a rectangular pulse, then for Δt we easily find a value equal to $\frac{\tau}{\sqrt{12}}=0.288\,\tau$. As for Δf, the integrals for M_{ω} and I_{ω} turn out to be divergent.

This is explained, generally speaking, by the fact that a rectangular pulse is characterized by a discontinuity in the function itself, and its spectrum falls off only as $\frac{1}{\omega}$. But the triangular pulse considered initially is characterized by a discontinuity in the first derivative, so that, consequently, its spectrum falls off as $\frac{1}{\omega^2}$; therefore the integrals prove to be convergent.

BIBLIOGRAPHY

[1] D. V. Ageev, "Active band of the frequency spectrum of a time function," Trudy Gor'kovskogo Politekhnicheskogo Instituta 11, 1, 5-10 (1955).

[2] N. N. Andreev, Electrical Oscillations and Their Spectra [in Russian] (Moscow, 1917).

[3] V. I. Vunimovich, Fluctuation Processes in Electronic Receivers [in Russian] (Soviet Radio Press, 1951).

[4] N. O. Bykova, "Action of a variable-frequency potential on resonance systems," Trudy MAP SSSR, No. 28 (1948).

[5] I. S. Gonorovskii, Electronic Signals and Transient Effects in Electronic Circuits [in Russian] (Svyaz'izdat, 1954).

[6] H. James, N. Nichols and R. Phillips, Theory of Tracking Systems [Russian translation] (IL, 1953).

[7] A. K. Kats, "Forced oscillations passing through resonance," Inzhenernyi Sbornik Instituta Mekhaniki Akad. Nauk SSSR 3, 2, 100-125 (1947).

[8] V. A. Kotel'nikov, "On the transmissiveness of the ether and wires in electrical communications," Materials for the First All-Union Conference [in Russian] (VEK, 1933).

[9] A. Yu. Lev and B. I. Yakhinson, "On the shifting of a signal spectrum," ES, No. 4, 68-74 (1956).

[10] J. Lawson and J. Uhlenbeck, Threshold Signals [Russian translation] (Soviet Radio Press, 1952).

[11] A. G. Maier and E. A. Leontovich, "On an inequality that is connected with the Fourier integral," Doklady Akad. Nauk SSSR 4, 7, 353-360 (1934).

[12] A. V. Rimskii-Korsakov, "On the analysis of oscillations," Trudy Komissi po Akustike Akad. Nauk SSSR, No. 6, 66-81 (1951).

[13] A. V. Rimskii-Korsakov and N. D. Shumova, "A method for the recording of frequency characteristics," Zhur. Tekh. Fiz. 8, 1478-1485 (1938).

[14] M. G. Serebrennikov, Harmonic Analysis [in Russian] (GTTI, 1948).

[15] I. T. Turbovich, "On the errors in measurements of frequency charac-

teristics by the method of frequency modulation," Radiotekhnika 9, 2, 31-35 (1954).

[16] I. Turbovich, "On the problem of dynamic frequency characteristics," Radiotekhnika (in press).

[17] B. B. Shtein and G. G. Varganov, "Taking out one of the side bands by means of multi-phase modulation," NT Sbornik MEIS, 13-24 (1950).

[18] Ch'iang Hsüeh-sên, Technical Cybernetics [Russian translation] (IL, 1956).

[19] N. F. Barber and F. Ursell, "The response of a resonant system to a gliding tone," Phil. Mag. 39, 345-361, 1948.

[20] P. Blassel, Erreur due a une durée d'intégration finie dans la détermination des functions d'autocorrélation," Ann. d. Télécom. 8, No. 12, 406-414, 1953.

[21] W. B. Davenport, R. A. Johnson and D. Middleton. "Statistical errors in measurements on random time functions," J. Appl. Phys. 23, No. 4, 377-388, 1952.

[22] R. M. Fano, "Short time autocorrelation functions and power spectra," 22, No. 5, 546-550, 1950.

[23] G. Hoke, "Response of linear resonant systems to excitation of a frequency varying linearly with time," J. Appl. Phys. 19, No. 3, 242-250, 1948.

[24] D. G. Lampard, "Generalization of the Wiener-Khintchin theorem to nonstationary processes," J. Appl. Phys. 25, No. 6, 802-803, 1954.

[25] D. G. Lampard, "Definitions of 'bandwidth' and 'time duration' of signals which are connected by an identity," Trans. IRE, CT-3 4, 286-288, 1956.

[26] E. M. Lewis, "Vibration during acceleration through a critical speed," Trans. Am. Soc. Mech. Eng. 54, 24, 253-261, 1932.

[27] C. H. Page, "Instantaneous power spectra," J. Appl. Phys. 23, 1, 103-106, 1952.

[28] R. K. Potter, G. A. Kopp and H. C. Green, Visible speech, (Van Nostrand, N. Y., 1947).

[29] S. O. Rice, "Filtered thermal noise-fluctuation of energy as a function of interval length," 14, No. 4, 216-227, 1943.

[30] S. O. Rice, "Mathematical analysis of random noise," Bell System Tech. J. 23, 3, 282-332 (1944); 24, 1, 46-156 (1945). Russian translation in the collection, Theory of the Transmission of Electrical Signals Past Obstacles (IL, 1953).

[31] L. M. Spetner, "Errors in power spectra due to finite sample," J. Appl. Phys. 25, No. 5, 653-659, 1954.

[32] L. A. Zadeh, "Correlation functions and spectra of phase- and delay-modulated signals," PIRE 39, No. 4, 425-427, 1951.